STP 1378

DISCARD

Hot Mix Asphalt Construction: Certification and Accreditation Programs

Scott Shuler and James S. Moulthrop, editors

ASTM Stock #: STP 1378

ASTM
100 Barr Harbor Drive
West Conshohocken, PA 19428-2959

Library of Congress Cataloging-in-Publication Data

Hot mix asphalt construction : certification and accreditation programs
/ Scott Shuler and James S. Moulthrop, editors.
    p. cm. -- (STP: 1378)
    ISBN 0-8031-2619-0
    1. Pavements, Asphalt.   2.   Industrial technicians--Certification-
-United States.   I.   Shuler, Scott.   II. Moulthrop, James S.,   1939
. III. Series: ASTM Special technical publication : 1378.
TE270.H68   1999
625.8'5--dc21   99-38880
                                                                 CIP

Peer Review Policy

Each paper published in this volume was evaluated by two peer reviewers and at least one editor.
The authors addressed all of the reviewers' comments to the satisfaction of both the technical
editor(s) and the ASTM Committee on Publications.

   To make technical information available as quickly as possible, the peer-reviewed papers in this
publication were prepared "camera-ready" as submitted by the authors.

   The quality of the papers in this publication reflects not only the obvious efforts of the authors and
the technical editor(s), but also the work of the peer reviewers. In keeping with long standing
publication practices, ASTM maintains the anonymity of the peer reviewers. The ASTM Committee on
Publications acknowledges with appreciation their dedication and contribution of time and effort on
behalf of ASTM.

Printed in Mayfield, PA
1999

Foreword

This publication, *Hot Mix Asphalt Construction: Certification and Accreditation Programs*, contains papers presented at the symposium of the same name held in Nashville, Tennessee, on December 8, 1998. The symposium was sponsored by ASTM Committee D-4 on Road and Paving Materials. Scott Shuler, Lafarge, Denver, Colorado and James S. Moulthrop, Koch Materials Company, Austin, Texas, presided as symposium Co-Chairmen and are the editors of the resulting publication.

Contents

Overview

Many construction processes are controlled by comparing a desired product, conceived during design, to the product produced during construction. The asphalt pavement construction process is often controlled in this manner. Control is often based on testing of components and assemblies of materials, the construction process, and the finished asphalt pavement. Success of the asphalt pavement construction project is usually judged based on how well test results produced during construction compare with criteria considered representing the desired product conceived during design.

Since success of an asphalt paving project is judged based on test results, it is logical that individuals conducting these tests be provided whatever training is necessary to assure the tests are conducted properly. The importance of this cannot be overemphasized. Significant sums of money depend on tests properly conducted. These sums of money represent not only the initial cost of the project, including payments to the contractor and subcontractors, but more significantly, performance of the pavement. Therefore, if test results do not reflect accurately true values of criteria representing pavement behavior, performance of the pavement may be in jeopardy.

There are at least three important components to consider when developing a process to control pavement construction. First, tests, which determine compliance with specifications, must be standardized. In asphalt pavement construction in the U.S., the process of developing and standardizing these tests is an ongoing process within bodies such as ASTM and AASHTO. Second, apparatus necessary to conduct the tests must be evaluated to determine competency. Third, the capability of personnel conducting the tests must be judged.

Statistical quality control and quality assurance (QC/QA) programs in hot mix asphalt pavement construction have become a significant contributor to more consistent and higher quality products. The result has been a steady improvement in the performance of asphalt pavements. The success of any QC/QA program is directly related to the quality of the data generated by technicians conducting the tests. Although standard test procedures published by ASTM and other bodies are used, differences in test results can still occur between the QC and QA laboratories. Reducing the potential for these differences is important so that an accurate estimate for the true value of each test result can be determined. Having confidence in these test results is important for controlling the manufacturing process. Certification and accreditation programs for both asphalt technicians and laboratories have been and are continuing to be developed to improve the consistency and quality of laboratory test results on asphalt paving construction projects. Successful programs accomplish this goal and provide additional benefits in the form of improved cooperation between the owner and contractor. The result is improved paving quality leading to increased performance in turn producing cost savings over the pavement life cycle.

Many states, municipalities, and other organizations responsible for asphalt paving have adopted various forms of certification programs for asphalt technicians and laboratories. Many more organizations intend to establish such programs in the near future either because the need has been clear or in response to FHWA, which has mandated certification programs by June 29, 2000 on federal aid projects as described in 23 CFR, Part 637.

Purpose of Symposium

This volume has been assembled to share the experiences of an assortment of organizations that have established or begun to establish programs for certification and accreditation for technicians working in the asphalt pavement construction industry. This information should be useful not only to those wishing to start new programs, but also to organizations with existing programs desiring to make improvements.

Our intent was to assemble as wide a variety of certification and accreditation programs from around the U.S. as possible. We hope those wishing to establish successful certification programs of their own can find helpful examples in the approaches presented.

Summary

It will become clear when reading this volume that a wide range of approaches has been taken when developing technician certification and accreditation programs around the country. A diverse group of organizations' experience has been compiled by the editors of this volume including departments of transportation, a trade association, a college and several universities, a paving contractor, and AASHTO. However, in spite of differences, much commonality can be identified between programs.

Perhaps the most ambitious program presented is described in the paper "New England Transportation Technician Certification Program (NETTCP): A Regional Approach." This program is a cooperative arrangement between six states that have agreed on the methods utilized to certify technicians so that an individual may work in any of the participating states. Training is an element in this program, which includes asphalt plant and laydown activities, aggregates, soils and portland concrete. In addition to certification, the program also is developing standardized test procedures to be followed in each of the six participating states.

"Asphalt Technician Certification: The Rocky Mountain Way" describes a program developed as a partnership between the Colorado DOT and the Colorado Asphalt Pavement Association. This partnership used seed money provided by CDOT and full-time personnel and dedicated laboratory space provided by CAPA, which operates and administers the program, to develop a combined classroom instruction and practical laboratory program to allow technicians to demonstrate proficiency. Five levels of certification are available: A—Laydown, B—Plant, C—Mixture Design, D—Profilograph, and under development, and E—Aggregates.

The AASHTO laboratory accreditation program is presented in the paper "The AASHTO Accreditation Program: Serving the Hot Mix Asphalt Industry." This paper summarizes the program beginning in 1988, which has accredited over 200 asphalt laboratories. The program requires that laboratories satisfy many quality systems and participate in the AASHTO Materials Reference Laboratory (AMRL) on-site assessment and proficiency sample program. The paper describes improved repeatability and reproducibility for accredited laboratories compared with the non-accredited.

"Experiences With Bituminous Paving Technician Training and Certification in Pennsylvania" describes a program which combines classroom and laboratory training with certification based on written exams. Three levels of asphalt plant technician are offered and one laydown technician. The program is primarily designed to be instructional with three- to four-day schedules for the various levels of certification. Although proficiency of technicians is not demonstrated in the laboratory, the program managers believe the close supervision provided during instruction is a good substitute. However, the paper states that practical proficiency demonstrations would be desirable if the scope of the program is expanded to accommodate it.

Arkansas DOT developed their program with the University of Arkansas described in the paper "A First-Year Summary of the Arkansas Hot Mix Asphalt Technician Certification Program." This program offers certification and training in aggregate technology, hot mix asphalt, portland concrete and soils. Written examinations are combined with practical laboratory evaluations of technician skill in conducting the various tests. Instructors are university staff, which has created some difficulty in scheduling the courses since summer is the best time for staff to conduct courses but the poorest time for prospective attendees.

A view of technician certification by a contractor is presented in the paper "Certification and Accreditation Programs: A Contractor's Perspective." The focus of this paper is to identify salient features of such programs such as written examinations, hands-on laboratory proficiency and laboratory accreditation. The message is to take advantage of the opportunity to standardize the technician certification and laboratory accreditation processes throughout the country so that different programs are not created in each state.

The only national certification program for asphalt technicians was developed starting in 1976 by the National Institute for Certification in Engineering Technologies (NICET). The program is described in the paper "Basic Elements in the Design of a Certification Program for Hot Mix Asphalt Construction Personnel." This paper describes the NICET model for certification which includes four elements: 1) acceptable completion of a written examination, 2) verification of practical competency by a direct supervisor, 3) satisfying the minimum work experience, and 4) satisfying a personal recommendation requirement. Some benefits of this program include third-party evaluation of strengths and weaknesses based on a standardized written examination, and a relatively rapid, economical program.

A community college in Illinois in partnership with two asphalt concrete producers provides certification for technicians in accordance with Illinois DOT requirements. The paper "Lake Land College/Illinois Department of Transportation: Quality Control/Quality Assurance Training Program—Development and Implementation" describes a program which includes certification in aggregates, three levels of hot mix asphalt, portland concrete, Superpave field control, and nuclear density testing. The program operates two laboratories located in strategic areas of the state to make it more convenient for attendees. Course length varies from two to five days.

"South Carolina's Experience With Certification and Accreditation" is a paper that describes five certifications available ranging from QC/QA laboratory and field personnel to mixture design and laboratory managers. University personnel administer the program but teams consisting of university, industry, and DOT experts teach classes. Written exams are part of all five courses and practical demonstrations of proficiency are required for two courses. Courses are limited to approximately 12 attendees and range from three to five days in length.

Closure

As more owners become aware of the benefits of quality control and quality assurance, the advantages of technician certification and laboratory accreditation will simultaneously become apparent. The intention of this volume is to present an assortment of certification and accreditation programs which measure the ability of personnel and the competency of the apparatus engaged in quality control and assurance testing. It is the editors' hope in assembling this volume that information provided here would be useful to practitioners wishing to establish new programs or improve existing programs by viewing the experience of others.

Scott Shuler

Lafarge
    1400 W. 65th Ave., Denver, Colorado, 80221;
    Symposium Co-Chairman and Editor

James S. Moulthrop

Koch Materials Company
    Austin, Texas, 78701;
    Symposium Co-Chairman and Editor

Leo C. Stevens,[1] Thomas Peterson,[2] and Christopher Bowker[3]

New England Transportation Technician Certification Program (NETTCP): A Regional Approach

---

Reference: Stevens, L. C., Peterson, T., and Bowker, C., **"New England Transportation Technician Certification Program (NETTCP): A Regional Approach,"** *Hot Mix Asphalt Construction: Certification and Accreditation Programs, ASTM STP 1378*, S. Shuler, and J. S. Moulthrop, Eds., American Society for Testing and Materials, West Conshohocken, PA, 1999.

Abstract: The six New England states have implemented a technicians certification program. Working together with the Federal Highway Administration (FHWA), the Federal Aviation Administration (FAA), the highway industry, academia, and private consultants throughout New England and New York, a non-profit organization was established entitled the New England Transportation Technician Certification Program (NETTCP). The intent of the program is to jointly develop training and certification courses that are supported and recognized by the New England states. Training and certification courses have been developed in a number of technical areas, including hot mix asphalt plant and hot mix asphalt paving, portland cement concrete and soils and aggregate. The background for this effort is the movement of the state Department of Transportations (DOTs) toward Quality Control/Quality Acceptance (QC/QA) specifications with the active support of the FHWA. In addition to the training and certification courses developed, there is movement to standardize the test methods which are used, i.e. American Association of State Highway and Transportation Officials (AASHTO) and/or ASTM, standardize test forms and to reduce problems with test result differences. The results of these efforts lead to a regional approach wherein technicians will be recognized as qualified in all six New England states. Future endeavors include additional positions at the technician level and the possible creation of a second level of certification that would be QC/QA technologists with specialties in hot mix asphalt, portland cement concrete and soils and aggregate.

Keywords: certification, training, technicians

---

[1] Self Employed, 15 Duxbury Road, Wellesley Hills, MA 02481.
[2] Executive Director, Colorado Asphalt Pavement Association, 6880 S. Yosemite Ct., Suite 110, Englewood, CO 80112.
[3] Executive Director, NETTCP, P.O. Box 722, Marshfield, MA 02050.

1

Introduction

The six New England States of Connecticut, Maine, Massachusetts, New Hampshire, Rhode Island and Vermont have implemented a technician certification program. Working together with the Federal Highway Administration (FHWA), the Federal Aviation Administration (FAA) and industry throughout New England, a nonprofit organization has been established named the New England Transportation Technician Certification Program (NETTCP). The intent of the program is to jointly develop training and certification courses that are supported by each of the New England States and that are commonly specified by each of them. Training and certification courses have been developed at the technician level for areas including hot mix asphalt, soils/aggregate and concrete. At this time, certification by NETTCP is being required by several New England States and is recognized by all New England States.

Overview

The NETTCP is a nonprofit organization with by-laws and is incorporated in the Commonwealth of Massachusetts. NETTCP consists of a twenty seven member Board of Directors composed of the six New England States Department of Transportation (DOT) Materials Engineers, representatives from FHWA and FAA, members of the contracting industry, materials suppliers, the private consultants and academia. A seven member executive committee is formed consisting of the officers, two elected members at large and the executive director.

There are four committees, which deal with particular subject areas for certification. They are Hot Mix Asphalt (HMA) Plant Technician, HMA Paving Inspector, Soils and Aggregate Technician, and Concrete Technician. Other committees presently active are Quality Control/Quality Acceptance (QC/QA) Technologist, Standardization, and Reciprocity.

Objectives

The objectives of NETTCP are:
1. To increase the knowledge of production and field technicians.
Through certification, minimum levels or benchmark levels of qualifications are established for both industry and agency personnel. Hopefully, the number of qualified technicians can be increased at the same time.
2. To reduce the problems associated with test result differences.
There will always be disputes because of test result differences. However, with certified technicians that put both sides on equal footing, the arguments won't be focused on who did it wrong or who's equipment is bad, but on other sources of variability, namely, the variability in the material.
3. To eliminate the issue of reciprocity of having individual state certification programs.
The regional or multi-state approach to certification eliminates the issue of reciprocity and allows contractors who cross state lines the freedom of not having to be recertified by

different programs. This is very important in New England where a significant number of contractors and material suppliers do work in several states.

4. To move forward in standardization of test methods and test procedures used by the six New England States.

Currently, there is a lack of uniformity throughout New England in what material test methods are used and the procedures used to perform the various American Association of State Highway and Transportation Officials (AASHTO) tests. The contractors and material suppliers that cross state lines would greatly benefit by any improvements of increased uniformity in this area.

5. To assist the New England States and industry in meeting the requirements of Federal Regulation 23 CFR Part 637, Construction and Materials.

As states move toward the QC/QA method of construction, there needs to be guidelines on what is a qualified technician. Setting up standards and qualifications for qualified technicians fill a need that is required.

Organizational Setup

As previously mentioned NETTCP has a set of by-laws and in addition publishes a Registration, Policies and Procedures Manual [1], the latest edition was printed in October 1998. This manual lists the membership of the Board of Directors, the Executive Committee, and the four Certification Committees. It establishes the certification courses, costs, details and requirements for each course and certification covered by NETTCP. It also lists those test methods for which candidates must exhibit written and performance proficiency. The manual prints the refund policy, the cancellation policy, the recertification policy and decertification policy as approved by the Board of Directors. It also has a procedure for complaints and/or protests. Membership in NETTCP is outlined in the by-laws of the organization.

The development and implementation of the NETTCP has been made possible through funding contributions from the FHWA, the FAA, the six New England states, contractors, consulting firms, and industry associations. The funding has allowed NETTCP to develop courses, retain the services of consultants to develop manuals, conduct pilot courses, work on standardization issues, develop and print certification, policies and procedures manual. The six New England states and the FHWA have collectively committed nearly $240,000 to the initial development of the program. Industry as a whole has contributed approximately $60,000. NETTCP continues to seek industry support through sponsorship and membership fees from industry. The primary use of these funds is development costs as the course fees are expected to cover the cost of the courses.

Certification Courses

Generally when a candidate registers for a course and examination, he or she must be a high school graduate, possess a GED, or receive prior approval from NETTCP. Prior approval consists of examining the resume of the candidate and following up on references provided in the resume. In addition work experience (using the example of

Hot Mix Asphalt Plant Technician) must consist of the following: a), work directly under a NETTCP Certified Hot Mix Asphalt Plant Technician for a minimum of 30 working days to demonstrate experience and proficiency in the test procedures outlined in the course and following the 30-day minimum work experience period, the candidate must be signed off by two NETTCP Certified Hot Mix Asphalt Plant Technicians indicating that the required work experience has been successfully completed or b), a candidate with two years verifiable relevant experience can be signed off by two NETTCP Certified Hot Mix Asphalt Technicians once the candidate demonstrates proficiency and knowledge with the required test procedures. To be certified a candidate must attend the certification course in its entirety and successfully pass the written and laboratory examination. The written examination consists of 60 questions, which can be multiple choice, true or false, or calculation. The candidate must achieve 70% to pass. The performance consists of three test procedures of which the candidate must achieve a minimum of 90% in each test procedure. A checklist is provided the examiner to provide guidance in setting test scores.

The Hot Mix Asphalt Plant Technician Certification Course is designed for those individuals responsible for the sampling and testing of hot mix asphalt at the production facility. The program is aimed at contractor and supplier quality control technicians, consultant testing firms and DOT inspectors responsible for the quality of hot mix asphalt. Currently the cost for the course and examination is $750.00 for members of NETTCP and $975.00 for non-members and consists of a four-day classroom and laboratory course. There is a one-day written and laboratory examination. Class size is 12 to 16 persons. The outline of study is shown in Table 1. Recertification costs $350.00 for members of NETTCP and $495.00 for non-members and consists of one-and-a-half-day classroom and laboratory courses with a written and performance examination.

The history to date is that the course structure and training manual for the HMA Plant Technician was developed in 1995 with the pilot presentation given in December 1995. In 1996, ten courses were presented and 139 technicians certified, in 1997, three courses were presented and 47 technicians certified and in 1998, four courses were presented and 42 technicians certified. In October 1998 a revised manual was made available for the 1999 courses and the recertification courses. The new manual will stress Superpave procedures but still retains some of the Marshall procedures.

The Hot Mix Asphalt Paving Inspector Certification Course is designed for those individuals responsible for the inspecting, sampling and testing of the hot mix asphalt in the field. The program is aimed at contractor and supplier quality control inspectors, consultant testing firms and DOT inspectors responsible for the quality assurance and placement of hot mix asphalt. Currently the cost for the course and examination is $375.00 for members and $495.00 for non-members and consists of a two-and-a-half-day classroom course and a two-hour written examination. Class size is 25 to 35 persons. The outline of study is shown in Table 2. In addition to the basic prerequisites for taking the course, candidates must also document current nuclear density gauge certification as required by the Nuclear Regulatory Commission or appropriate state nuclear regulatory agency. The written examination consists of 60 questions, which can be multiple choice, true or false, or calculation. The candidate must achieve 70% to pass.

The history to date is that the course structure and training manual for the HMA Paving Inspector was developed in 1996 with the pilot presentation given in January 1997.

Table 1 - *Outline Of Study For Hot Mix Asphalt Plant Technician*

ASTM/ AASHTO	Procedure	Previous Experience Required	Test Written (W) Physical (P)
T 2	Sampling of Aggregates	Yes	W
T 248	Reducing Field Samples of Aggregates to Testing Size	Yes	W
	Hot Bin Samples or Cold Feed Samples	Yes	W
T 27	Sieve Analysis of Fine and Coarse Aggregate	Yes	W
T 11	Material Finer than the 75μm (No. 200) Sieve by Washing		W
T 168	Sampling Bituminous Paving Mixtures	Yes	W
T164 or TP 53	Quantitative Centrifuge Extraction of Ignition Oven	Yes	W
T 30	Mechanical Analysis of Extracted Aggregate	Yes	W
TP 4	Fabrication and Testing of Superpave Paving Mixture Specimens	Yes	W,P
	Mix, Asphalt, Air Temperature	Yes	W
T 245	Resistance to Plastic Flow of Bituminous Mixtures Using Marshall Apparatus	Yes	W,P(New certifications)
T 40	Sampling Bituminous Materials		W
	Consensus Aggregate Tests		W
T 84	Specific Gravity and Absorption of Fine Aggregate		W
T 85	Specific Gravity and Absorption of Coarse Aggregate		W
T 166	Bulk Specific Gravity of Compacted Bituminous Mixtures using Saturated-Surface Dry Specimens		W,P
T 269	Air Void/V.M.A./V.F.A. Calculations		W
	Combined Hot Bin Analysis		W
T 255	Total Moisture Content of Aggregate by Drying		W
T 209	Maximum Specific Gravity of Bituminous Paving Mixtures		W,P
T 269	Thickness of Compacted Paving Specimens		W,P
	Preparations of Cores for TMD		W
ASTM D 3665	Random Sampling		W

This course uses The Asphalt Institute Manual MS-22, the NAPA Handbook, as well as other selected material. In 1997, five courses were presented and 154 technicians certified, in 1998, six courses were presented and 155 technicians certified.

Table 2 - *Outline Of Study For Hot Mix Asphalt Paving Inspector*

AASHTO/ ASTM	Procedure	Previous Experience Required	Test Written (W) Performance (P)
	Materials and HMA Mixture		W
	Production Facilities		W
	Surface Preparation	Yes	W
	Mixture Delivery and Placement	Yes	W
	Joint Construction	Yes	W
	Compacting the Mat	Yes	W
	Quality Control, Quality Assurance		W
ASTM D 3665	Random Sampling		W
T 168	Field Sampling	Yes	W
	Mat Troubleshooting		W
T 166	Bulk Specific Gravity of Compacted Bituminous Mixtures Using Saturated Surface Dry Specimens	Yes	W
T 209	Maximum Specific Gravity of Bituminous Paving Mixtures		W
T 269	Thickness of Compacted Paving Specimens		W
ASTM D 2950	Density of Bituminous Concrete in Place by Nuclear Methods	Yes	W

The Soils and Aggregate Technician Certification Course is designed for those individuals responsible for the sampling and testing of soils and aggregates used in base, subbase and roadway embankment construction. The program is aimed at contractor and supplier quality control technicians, consultant testing firms and DOT inspectors responsible for the quality assurance and placement of select aggregate and soil materials. Currently the cost for the course and examination is $475.00 for members and $650.00 for non-members and consists of a two-and-a-half-day classroom and laboratory course. There is a one-half-day written and laboratory examination. Class size is 12 to 16 persons. The outline of study is shown as Table 3. In addition to the basic prerequisites for taking the course, candidates most also document current nuclear density gauge certification as required by the Nuclear Regulatory Commission or appropriate state nuclear regulatory agency. The written examination consists of 70 questions, which can be multiple choice, true or false, or calculation. The candidate must achieve a minimum score of 70% to pass.

The performance test consists of three test procedures of which the candidate needs to achieve a minimum of 90% in each test procedure.

Table 3 – *Outline of Study for Soils and Aggregate Technician*

AASHTO / ASTM	Procedure	Previous Experience Required	Test Written (W) Performance (P)
T 2	Sampling Aggregates	Yes	W
ASTM D 3665	Random Sampling		W
T 248	Reducing Field Sample of Aggregates to Testing Size	Yes	W,P
T 27	Sieve Analysis for Fine and Coarse Aggregate	Yes	W,P
T 11	Material Finer than 75µm (No. 200) Sieve by Washing	Yes	W,P
T 104	Soundness of Aggregate - Sodium or Magnesium Sulfate		W
T 96	Resistance to Abrasion/LA Wear Test		W
T 88	Particle Size Analysis of Soils		
T 84	Specific Gravity & Absorption of Fine Aggregates		W
T 85	Specific Gravity & Absorption of Coarse Aggregate		W
T 255	Moisture in Aggregate by Drying	Yes	W
T 217	Moisture in Aggregate by Gas Pressure Method		
T 99 & T 180	Moisture Density Relation of Soils	Yes (Either One)	W,P
T 191	Density of Soil by Sand-Cone Method	Yes	W
T 238	Density of Soils and Soil Aggregate in Place by Nuclear Method	Yes	W
T 224	Correction for Stone in Compaction Tests		W
T 89	Liquid Limit of Soils		W
T 90	Plastic Limit & Plasticity Index of Soils		W
ASTM D 2488	Identification of Soils by Visual Procedure		W,P

   The history to date is that the course structure and training manual for the Soils and Aggregate Technician was developed in 1996 with the pilot presentation given in December 1996. It should be noted that the New Jersey DOT participated in developing this course and manual. In 1997, four courses were presented and 58 technicians certified and in 1998, six courses were presented and 67 technicians certified.

The Concrete Technician Certification Course is designed for those individuals responsible for sampling and testing of portland cement concrete and related materials at either the plant or in the field. The program is aimed at contractor and supplier quality control inspectors, consultant testing firms and DOT inspectors responsible for the quality assurance of portland cement concrete. Currently the cost for the course and examination is $140.00 for members and $225.00 for non-members and consists of a one-day classroom course and a one-hour written examination. Class size is 25 to 35 persons. The outline of study is shown in Table 4. In addition to the basic prerequisites for taking the course, candidates also must document current ACI Grade I Field Certification. The written examination consists of 50 questions, which can be multiple choice, true or false, or calculation. The candidate must achieve a minimum mark of 70% to pass.

Table 4 – *Outline of Study for Concrete Technician*

AASHTO /ASTM	Procedure	Previous Experience Required	Test Written (W) Performance (P)
T 2	Sampling Aggregates	Yes	W
ASTM D 3665	Random Sampling		W
T 248	Reducing Field Sample of Aggregates to Testing Size	Yes	W
T 27	Sieve Analysis for Fine and Coarse Aggregate	Yes	W
T 255	Moisture in Aggregate by Drying	Yes	W
	Aggregate Blending		W
	Cement, Pozzolans & Admixtures		W
	Concrete Plant Inspections		W
	Batch Weight Adjustment		W
	Properties of Concrete Field Tests		W

The history to date is that the course structure and training manual for the Concrete Technician was developed in 1997 with the pilot presentation given in December 1997. In 1998, three courses were presented and 83 technicians certified.

QC/QA – Quality Assurance

With the trend towards more projects being bid as QC/QA contracts by many states, and with guidance from the FHWA, NETTCP has undertaken this task with the outlook as to how it can aid the New England states in the adoption of QC/QA specifications. Under Quality Assurance, the Contractor is responsible for Quality Control (QC) and the state DOT is responsible for Quality Acceptance (QA). If the state desires, testing by the contractor could be also used for Quality Acceptance. There also is Independent Assurance Testing conducted by the state DOT. Contractors will be required by the specifications to submit a Quality Control Plan to the state DOT. In this plan the contractor must spell out who will be the qualified personnel responsible for the quality control testing, the qualified testing facility that will be used, the Plan Administrator, and the quality control sampling and testing program he will be using. The state DOT, who will be doing the quality acceptance testing, must have qualified personnel to perform their acceptance testing, must have a qualified laboratory and be responsible for providing a plan for acceptance testing. Independent assurance testing is a check upon the technicians knowledge on how to perform the test in accordance with AASHTO and/or ASTM test procedures and if the equipment being used meets AASHTO and/or ASTM test procedures. Current QC/QA specifications in some of the New England states require that the qualified technicians be NETTCP certified.

Courses and examinations are held in various locations throughout New England. Those courses, which do not have a proficiency component such as the Hot Mix Asphalt Pavement Inspector Program, typically will be held in a hotel or other facility with a meeting room capable of holding fifty persons. Those programs with a proficiency component will be held at a materials suppliers plant laboratory or a DOT laboratory with a meeting room nearby.

Developmental Program

NETTCP began looking to the future with a Developmental Program dated October 1997. This program looked into where we are going once the basic technician certification programs matured and are self-sustaining. The plan envisioned a three-tier system of certifying technicians, technologists and plan administrators. Level I - consists of the technician group and the inspector group. The technician group is centered primarily on sampling, testing and basic inspection. These courses required as a prerequisite a combination of education and testing experience as per NETTCP policy (ACI Certification required for Concrete Technician). These included the HMA Plant Technician, Soils & Aggregate Technician, Concrete Technician, and under serious discussion presently is a HMA binder technician. The inspector group includes HMA Paving Inspector, Concrete Inspector (under development), and Soils & Aggregate Inspector (to be developed later). Other efforts may include a paint technician category. These will require as a course - requisite education and inspection experience as per NETTCP policy manual. Level II - the QC/QA Technologist level is centered primarily on understanding QC/QA philosophy, calculations of Percent Within Limits (PWL), etc. and consists of three modules. The first modules are: Module I Quality Assurance Specifications (QC/QA) - An Overview, Module II Quality Control and Module III Statistical Concepts. Level III - Plan Administrator is still in the future.

Other Activities

NETTCP has two other activities, the first is standard test methods, and the second is standard test reports. A survey was taken of the six New England States to determine which test method was used, AASHTO, ASTM, or other. Basically, most tests were AASHTO, some with variation, some were ASTM and some were state-written procedures. The biggest area of difference was for portland cement concrete, some states used ASTM and some used AASHTO. The testing procedures committee is slowly working on getting the states to agree to use the same procedures such as AASHTO. The initial efforts are with aggregate test procedures.

The standard test report committee has completed initial efforts to create a standard test form for tests required under certification. A contract has been awarded to a firm to provide a finished product in Microsoft Excel. Another activity is a committee working with New York State on reciprocity with regards to HMA Plant Technician. This is a time-consuming effort, which is probably reflected in other sections of the country.

Reference

[1] New England Transportation Technicians Certification Program, "Procedures, Policies and Registration Program," NETTCP, P.O. Box 722, Marshfield, MA 02050, 1998.

Michael M. Cassidy[1] and Scott A. Conner[2]

Asphalt Technician Certification: The Rocky Mountain Way

Reference: Cassidy, M. M., and Conner, S. A., **"Asphalt Technician Certification: The Rocky Mountain Way,"** *Hot Mix Asphalt Construction: Certification and Accreditation Programs, ASTM STP 1378,* S. Shuler, and J. S. Moulthrop, Eds., American Society for Testing and Materials, West Conshohocken, PA, 1999.

Abstract: The Asphalt Technician Certification Program in Colorado is one of four programs offered by the Rocky Mountain Asphalt Education Center located in Englewood, Colorado. Established in 1996, the Laboratory for Certification of Asphalt Technicians (LabCAT) is operated by the Colorado Asphalt Pavement Association in a partnership effort with the Colorado Department of Transportation (CDOT) and the Federal Highway Administration. The LabCAT was developed to increase the proficiency of testing technicians and respond to the Federal requirements of having qualified technicians performing sampling and testing on Federal-Aid Projects. The objective of the LabCAT is to certify technicians directly involved with identifying the properties of the final asphalt product in any Quality Control/Quality Assurance (QC/QA) program.

    Four certification levels have been established for Asphalt Construction Technicians based on material tests on typical paving projects. The certification is valid for a three-year period. The certification levels are: Level A-Laydown; Level AB-Laydown and Asphalt Plant Materials Control; Level C-Mixture Volumetrics and Hveem Stability; and Level D-Smoothness. A new level of certification is currently under development. Titled "Level E-Aggregate Technician Certification," this level will have the same structure as the existing program levels but will be based on the Superpave aggregate testing protocols.

    As of November 1998, 419 Asphalt Construction Technicians have attended one or more of the certification levels at the LabCAT. Initially, the certifications were based upon the standard written Colorado Department of Transportation Procedures (CP or CP-Laboratory). In anticipation of regional reciprocity in asphalt technician certification and certification of City and County personnel, in 1998 the program was modified to certify technicians based upon the AASHTO *Standard Specifications for Transportation Materials and Methods of Sampling and Testing* with appropriate references to the CDOT *Field Materials Manual* and the CDOT *Laboratory Manual of Test Procedures*.

Keywords: hot mix asphalt, asphalt technician certification, certification programs

[1] Laboratory Manager, Rocky Mountain Asphalt Education Center, Englewood, CO.
[2] Assistant Lab Manager, Rocky Mountain Asphalt Education Center, Englewood, CO.

Introduction

The Asphalt Technician Certification Program in Colorado is one of four programs offered by the Rocky Mountain Asphalt Education Center (RMAEC) located in Englewood, Colorado. Operated as a separate entity, the Laboratory for Certification of Asphalt Technicians (LabCAT) was established in 1996 to increase the proficiency of the asphalt technicians, improve the reliability of Quality Control and Quality Assurance (QC/QA) testing and increase the quality of asphalt paving materials purchased by owner-agencies. The objective of the certification program is to certify those individuals directly responsible for identifying the properties of the final asphalt product in any QC/QA program and therefore the quality level. The LabCAT program is a partnership between the Colorado Department of Transportation (CDOT), the Colorado Asphalt Pavement Association (CAPA), and the Federal Highway Administration (FHWA).

The Rocky Mountain Asphalt Education Center (RMAEC) includes a fully-equipped asphalt material testing laboratory and is an AASHTO accredited facility.

Overview

The proposal for the Colorado Asphalt Construction Technician Certification was the result of a joint effort between CAPA, CDOT and FHWA. Principle contacts were Scott Shuler, the CAPA Executive Director; Dennis Donnelly the CDOT Research Engineer; and Doyt Bolling, the Regional Pavement/Materials Engineer, FHWA Region 8. The proposal was submitted to the Colorado Transportation Commission in August of 1994 for approval and a contract between CDOT and CAPA was finalized in July 1995.

Four certification levels were established for Asphalt Construction Technicians based on typical tests required for Quality Control and Quality Assurance programs and are listed in Table 1.

Table 1 - *LabCAT Levels of Certification*

Level	Certification
A	Laydown
AB	Laydown and Asphalt Plant Materials Control
C	Volumetrics and Stability
D	Smoothness

The Asphalt Construction Technician Certification Programs are stand-alone certifications, but are always offered sequentially so individuals have the opportunity to become certified in one or all of the programs. All certifications are based on the

American Association of State Highway and Transportation Officials *"Standard Specifications for Transportation Materials and Methods of Sampling and Testing,"* with the CDOT *Field Materials Manual* and the CDOT *Laboratory Manual of Test Procedures* referenced and incorporated into the course instruction, where appropriate.

Management Structure

The CAPA operates the LabCAT in a cooperative effort with the CDOT and the FHWA. The Board of Directors of the RMAEC/LabCAT is composed of three members each of CDOT and CAPA, and one representative of the FHWA. In order to broaden the scope and effectiveness of the educational and certification programs offered by the RMAEC, two additional members were added to the Board of Directors, a county representative and a city representative.

All asphalt technician certifications are conducted at the RMAEC/LabCAT which employs two full-time instructors, the Manager and Assistant Manager of the RMAEC. In addition to the Board of Directors, a technical group has been established to provide input into the requirements and scope of the certification and educational programs. The technical group is composed of the RMAEC manager, two CDOT regional materials engineers, an FHWA representative, one county representative, one contractor representative, and two consultant representatives.

The Format

The format for the Asphalt Technician Certification Program is composed of two parts. The first part is a brief classroom session of presentations on the basic principles of sampling, splitting and testing procedures of Hot Mix Asphalt (HMA) materials and the purpose of the tests. At the conclusion of the classroom session, the participants take an open book, written examination. The second part of the program consists of small group demonstrations of the sampling, splitting and testing procedures by the LabCAT staff. During this portion, the most critical aspects of each procedure are explained to the attendees. After the demonstrations, each participant is required to demonstrate proficiency, in a closed-book session, in each of the required procedures for certification to the LabCAT Proctor. To become certified, each participant must successfully pass the written exam(s) and all of the proficiency demonstrations required for certification.

The Programs

Level A - Laydown

Technicians responsible for sampling Hot Mix Asphalt (HMA) and HMA aggregates, and conducting compaction tests of HMA concrete at the laydown site are required to have this certification. The proficiency requirements for the Level A-Laydown Certification are listed in Table 2.

Table 2 - *Proficiency Requirements for Level A*

Procedure	AASHTO	CDOT
Practice for Stratified Random Sampling of Materials		CP - 75
Practice for Sampling Bituminous Paving Mixtures	T 168	CP - 41
Practice for Sampling Aggregates	T 2	CP - 30
Test Method for Reducing Field Samples of Hot-Mix Bituminous Pavements to Testing Size	T 248	CP - 55
Test Method for Density and Percent Relative Compaction of In-Place Bituminous Pavement by the Nuclear Method		CP - 81
Compaction Test Section, Coring, and Core Correlations		

Level AB - Laydown and Asphalt Plant Materials Control

Technicians responsible for sampling HMA and HMA aggregates, conducting compaction tests, and material process control at the HMA plant are required to have this certification. The proficiency requirements for the Level AB-Laydown and Asphalt Plant Materials Control Certification are shown in Table 3.

Table 3 - *Proficiency Requirements for Level AB*

Procedure	AASHTO	CDOT
Stratified Random Sampling of Materials		CP - 75
Sampling Bituminous Paving Mixtures	T 168	CP - 41
Sampling Aggregates	T 2	CP - 30
Reducing Samples of HMA to Testing Size	T 248	CP - 55
Coring and Handling		
Compaction Test Section		
Nuclear versus Core Correlations		
Reducing Samples of Aggregates to Testing Size	T 248	CP - 32
Sieve Analysis of Fine and Coarse Aggregates	T 27	CP - 31A

Table 3 continued

Table 3 - *Proficiency Requirements for Level AB - Continued*

Procedure	AASHTO	CDOT
Materials Finer Than 75-μm (No. 200) Sieve in Mineral Aggregates by Washing	T 11	CP - 31B
Test Method for Asphalt Cement Content of Asphalt Concrete Mixtures by the Nuclear Method	T 287	CP - 85
Bulk Specific Gravity of Compacted Bituminous Mixtures Using Saturated Surface-Dry Specimens	T 166	CP - 44
Theoretical Maximum Specific Gravity and Density of Bituminous Paving Mixtures	T 209	CP - 51
Control Chart Fabrication		

Level C - Volumetrics and Stability

Technicians responsible for determining mixture volumetric and Hveem stability characteristics for HMA produced at the HMA plant are required to have this certification. The proficiency requirements for Level C-Volumetrics and Hveem Stability Certification are given in Table 4.

Table 4 - *Proficiency Requirements for Level C*

Procedure	AASHTO	CDOT
Stratified Random Sampling of Materials		CP - 75
Sampling Bituminous Paving Mixtures	T 168	CP - 41
Sampling Aggregates	T 2	CP - 30
Reducing Samples of HMA to Testing Size	T 248	CP - 55
Reducing Samples of Aggregates to Testing Size	T 248	CP - 32
Asphalt Content by Nuclear Method	T 287	CP - 85
Bulk Specific Gravity of Compacted Specimens	T 166	CP - 44
Theoretical Maximum Specific Gravity of HMA	T 209	CP - 51
Control Chart Fabrication		

Table 4 continued

Table 4 - *Proficiency Requirements for Level C - Continued*

Procedure	AASHTO	CDOT
Test Method for Resistance to Deformation and Cohesion of Bituminous Mixtures by Means of Hveem Apparatus	T 246	CP -L5106
Method for Preparing and Determining the Density of Hot Mix Asphalt (HMA) Specimens by Means of the SHRP Gyratory Compaction	TP 4	CP - L5115
Practice for Volumetric Analysis of Compacted Hot Mix Asphalt (HMA)	PP 19	CP 48

Level D - Smoothness

Technicians responsible for measuring smoothness using the California rolling profilograph are required to have this certification by attending the program or by purchasing the videotape version produced by CAPA. The proficiency requirement for the Level D-Smoothness Certification is shown in Table 5.

Table 5 - Proficiency Requirement for Level D

Procedure	AASHTO	CDOT
Operation of Multi-Wheel Profilograph and Evaluation of Profiles, 0.1 inch (2.5 mm) (For Hot Bituminous Pavements)		CP - 70

Operations

All certifications take place at the Laboratory for Certification of Asphalt Technicians (LabCAT) classroom and laboratory located at 6880 South Yosemite Court, Suite 110, Englewood, Colorado. The LabCAT contains a classroom, with seating for up to 24 attendees, and a fully-equipped HMA testing facility. The facility consists of seven laboratory areas where the testing demonstrations and proficiency demonstrations are performed. Many of the required tests have multiple stations in different labs.

Program Length

The length of the program depends on the desired level(s) of certification. The Level A Certification is completed in one day. The Level AB Certification is usually completed in two days. The Level C Certification, or Level ABC Certification is typically completed within three days. The Level D Certification is a ½ - day course.

Program Materials

Classroom training material consists of instruction aids developed by the LabCAT and agreed upon by the program sponsoring agencies (CDOT and CAPA). Standard written Colorado Department of Transportation CP or CPL, or AASHTO laboratory sampling, splitting and testing procedures to be evaluated will be provided to each participant. In addition, literature published by the Asphalt Institute, the National Asphalt Pavement Association, Federal Highway Administration and others are used to supplement the training aids developed by the LabCAT.

Re-Certification

Each certification is valid for three years. If attendees take different levels of certification at separate times, the expiration dates of the levels will be different. Asphalt Construction Technicians have to attend the certification program every three years to sustain their certified status.

Attendance

The total number of Asphalt Construction Technicians who attended the Certification Program since 1996 is given in Table 6. It must be remembered that a technician could attend one or more levels of certification. That is, a technician who attended the Level A - Laydown is counted the same as a technician who attended Level ABCD - Laydown, Plant Materials Control, Volumetrics and Stability, and Smoothness. The technician who attended the Level ABCD Certification had to pass four written exams, and demonstrate proficiency in fifteen procedures, while the Level A technician had to pass one exam and demonstrated proficiency in eight procedures.

Table 6 - *Number and Percent Passing of Attendees to the Programs*

	1996	1997	1998 (to Nov.)	Total (to date)
Attendees, #	191	137	91	419
Passed at Initial Attempt, #	163	73	62	298
Percent Passed at Initial Attempt, %	85	53	74	71
Passing at First Retest, #	17	50	16	83
Percent Passed at First Retest, %	90	96	89	93

As can be seen from Table 6, the percent passing from 1996 to 1997 shows a decrease, with a rebound in the percent passing in 1998. There are two reasons to explain this.

First, the knowledge and experience of the initial attendees to the certification program were pronounced. When the second year began, there was a noticeable lack of knowledge and experience in the participants attempting to pass the certification program. The difference in testing experience is a major factor for the increase in the failure rate, and that leads to the second reason the percent passing decreased in 1997. When the certification program began in 1996, the laboratory proficiency demonstrations were an "open book" proficiency test. The technician could look up certain parts of the sampling, splitting and testing procedures, if necessary. The LabCAT staff reasoned that when performing laboratory testing, if a certain temperature or drying time could not be remembered, the technicians were taught to look it up in the procedures manual. In January 1997, the technicians seeking to become certified began to read the procedures directly from the laboratory testing manual to the Proctor when describing how to perform a test. It was at this point that technicians were required to be able to pass the proficiency demonstrations in a "closed book" environment. Since many of the participants did not seem to have the experience to perform the tests in this manner, the percent failing increased. As the word spread that the technicians seeking certification needed to know how to correctly perform the proficiency testing without looking in the procedures manual, the training for these individuals increased before they were sent to the LabCAT.

The number of Certified Asphalt Construction Technicians, the year of certification, and what level of certification was attained are given in Table 7.

Table 7 - *Number of LabCAT Certified Technicians by Level and Year*

Year	Level A	Level AB	Level ABC	Level ABCD	Level ABD	Total
1996	24	64	30	35	32	185
1997	18	73	12	11	11	125
1998	11	33	11	13	5	73
Total	53	170	53	59	48	383

Retests / Grading

Certification Requirements-Each participant must successfully pass the written examination(s) and each of the proficiency demonstrations in the laboratory. Participants to the program can request a retest if: a) the participant completed the entire program for the desired level of certification(s); and b) received a failing score on the written examination(s) or any of the proficiency demonstrations.

Written Examination-A passing grade requires a minimum of 80 percent correct answers on each of the written examinations. Tests can be retaken for scores below 80 percent, a minimum of two weeks following the first failure.

Laboratory Proficiency-Each of the levels of certification requires proficiency demonstrations in the laboratory, in a "closed book" environment. Detailed checksheets have been developed for use at the LabCAT based on current AASHTO and CDOT sampling, splitting and testing procedures for asphalt materials. A checksheet is used for each of the required tests for certification, with questions asked and critical steps of the procedure weighted by importance to the test. The maximum number of points for the proficiency testing ranges from 100 to 250 points. Each participant is allowed to miss 20 points per proficiency test. Tests can be retaken for participants who miss more than 20 points on a proficiency demonstration, a minimum of two weeks following the first failure. When the individual returns to be retested in the laboratory, in addition to the requirement to demonstrate proficiency in the test(s) the individual failed, they must successfully demonstrate proficiency in one test that they have previously passed.

Fees

The fees for each level of certification are determined by the LabCAT Board of Directors. Personnel attending the certification programs who work for the CDOT or for members of CAPA attend the program at a reduced rate. The cost of certification also depends on the level of certification desired. The average cost for CDOT/CAPA members is approximately $150/day, while the cost for nonmembers of CDOT or CAPA is approximately $240/day. Personnel attending the certification programs who work for cities or counties can do so at a reduced rate in recognition of the fact that cities and counties often have limited training/certification funds.

De-Certification

Certification is a privilege, and this privilege may be revoked if the individual is thought to have knowingly committed acts which are detrimental to the integrity of the certification program or the construction industry in general. Acts which could result in revocation of the certification privileges are:
- Falsification of field or quality control tests results and/or records.
- Cheating on certification exams.
- Submittal of false information on certification applications.
- Termination of an individual due to job incompetence.
- Criminal action by an individual while engaged in construction activities.

If, in the opinion of the LabCAT Board of Directors, revocation of certification privileges is warranted, an individual will receive a written notification stating such. The individual will be allowed sixty days within posting of the notification to respond by letter to the program administrator. If, during that time, a written letter of protest is received from the individual, the case will be reviewed by the LabCAT Board of Directors and the individual will be notified of a final decision. If no protest letter is received, it will be assumed by the program administrator that the individual does not protest the decision and revocation will occur with the individual so notified.

The Future

The objectives of the LabCAT are to increase the proficiency of the asphalt technicians, improve the reliability of QC/QA testing and increase the quality of asphalt paving materials. The feedback from participants in the program, and from the industry in general, is that the LabCAT is achieving these goals. Technicians are being trained prior to attempting to become certified. Through the LabCAT, all technicians on CDOT QC/QA projects must be certified. On projects where a certified QA technician has test results different from the certified QC technician, they will try to find the problem rather than accuse the other of not understanding the test procedure.

The LabCAT staff is continually changing, modifying and improving the certification program by updating testing procedures, by introducing new techniques and the latest technologies. The LabCAT was established at an opportune time to help implement the Superpave protocols. The volumetric element of Level-C is based on the theoretical maximum specific gravity and the bulk specific gravities of specimens compacted with the Superpave gyratory compactor. A three-year developmental plan has been approved to guide the way for the next three years.

When the LabCAT began certifying technicians, the levels of certification were based on CDOT standard methods and procedures. In 1998 however, the levels of certification were modified to follow the AASHTO standard methods and procedures for two reasons. First, to act as a resource for City and County Asphalt Construction Technicians who would not know the Department of Transportation's standard methods of sampling and testing, but are familiar with the AASHTO procedures. The second reason is a possible expansion of the LabCAT into a regional certification facility. With a curriculum based on AASHTO standard procedures and methods, states that do not have an Asphalt Construction Technician Certification Program could accept the LabCAT certification. In addition, the LabCAT could work with those states that do have an existing program to achieve reciprocity. Some of the barriers to a regional certification program include: out-of-state travel, state test methods versus AASHTO, and existing programs in states that may not want to participate in reciprocity.

Changes to the Program

Some changes are in the works for the LabCAT in 1999 and the future. One of these has been the addition of the Test Method for Resistance of Compacted Bituminous Mixtures to Moisture Induced Damage (T 283) as a requirement to the Level C - Volumetrics and Stability Certification. When the LabCAT was established in 1996, T 283 was not included in the CDOT QC/QA Program. It was decided to include T 283 in 1999 when CDOT added T 283 as an element of process control in projects constructed using voids acceptance protocols.

Another change that will be implemented in 1999 is the development and implementation of a new level of certification: Level E - Aggregate Technician Certification Program. This level of certification was established to ensure that testing on the aggregates used in HMA mixture designs are performed by qualified technicians.

The format of Level E Certification is similar to the levels already offered at the LabCAT and is composed of two parts. The first part is a classroom session of presentation on the basic principles of aggregate sampling, splitting and testing. At the conclusion of the session, a 40-minute written examination is given on the basics of these elements. The second part consists of small group demonstrations of the Superpave aggregate sampling and testing procedures. After the demonstrations, each participant is required to demonstrate proficiency in each test to the LabCAT Proctor. The grading for the Level E - Aggregate Technician Certification is the same as the other levels of certification. The test procedures included in the program are listed in Table 8. The *C* and *L* indicate the procedures that participants would be tested on during the written examination (C) and to demonstrate proficiency in the laboratory (L).

Table 8 - *Required Aggregate Tests for Certification*

Test	Method	Class	Lab
Sampling Aggregates	T2 / CP-30	C	L
Materials Finer than 75-μm, Sieve Analysis	T11, 27 / CP-31	C	
Specific Gravity and Absorption of Fine Aggregate	T84 / CP-L4102	C	L
Specific Gravity and Absorption of Coarse Aggregate	T85 / CP-L4103	C	L
Resistance to Degradation of Small-sized Coarse Aggregate by Abrasion and Impact in the Los Angeles Machine	T96	C	
Soundness of Aggregate by Use of Sodium Sulfate or Magnesium Sulfate	T104	C	
Clay Lumps and Friable Particles in Aggregate	T112	C	
Plastic Fines in Graded Aggregates and Soils by Use of the Sand Equivalent Test	T176	C	L
Reducing Samples of Aggregate to Testing Size	T248 / CP-32	C	
Uncompacted Void Content of Fine Aggregate	T304 / CP-L5113	C	L
Test Method for Determining Percent of Particles with Two or More Fractured Faces	CP-45	C	L
Test Method for Flat Particles, Elongated Particles, or Flat and Elongated Particles in Coarse Aggregate	D4791	C	
The 0.45 Power Curve			

A third change to the LabCAT Program for 1999 is the modification of the Asphalt Technician Certification Program. Many of the technicians who attended the certification program in 1996 will have to return in 1999. As stated earlier, there was a high number of experienced and knowledgeable participants during the first year the LabCAT was in operation. To accommodate these technicians, a "streamlined" certification program is being developed. The streamlined program will have the same requirements of certification as the present program, but classroom discussions and demonstrations of the sampling, splitting and testing procedures will be eliminated. The streamlined certification program will be a one-day program. The participants will attend a brief classroom session, take the written exam(s) and demonstrate proficiency for the sampling, splitting and testing procedures required for the desired level of certification.

Summary

The Laboratory for Certification of Asphalt Technicians (LabCAT) is the result of a combined effort with the CDOT, CAPA and FHWA. Established in 1996, the LabCAT has certified 383 Asphalt Construction Technicians in Colorado and the surrounding states. Four levels of certification are offered depending the responsibilities of the technician: Level A-Laydown, Level AB-Laydown and Plant Materials Control, Level C-Volumetrics and Stability, and Level D-Smoothness. A new level of certification, Level E-Aggregate Technician Certification will be offered in 1999.

The objective of the LabCAT is to increase the proficiency of asphalt technicians, improve the reliability of QC/QA testing, increase the quality of asphalt paving materials purchased by owner/agencies and respond to the Federal requirements of having qualified technicians performing sampling and testing on Federal-Aid Projects. With the implementation of QC/QA specifications, it is important to have qualified personnel performing the sampling, splitting and testing procedures to assure the quality of the material, reduce the problems associated with test result differences and to improve the confidence and level of expertise of the Asphalt Construction Technician.

The LabCAT is working continually to improve the course curriculum by introducing new technologies and test methods, and updating the course presentations and manuals to include the latest information available. A 3-year developmental plan has been approved and implemented to ensure that the work is accomplished on a timely basis and that we move forward.

There are a number of challenges to truly becoming a regional program. However, the LabCAT is in a good position to act as a regional facility, whether it is with cities, counties or states. There are several hurdles to overcome prior to the full implementation of a regional certification program. These hurdles include: a) out-of-state travel; b) long distances between Denver and other western locations; c) different sampling, splitting and testing procedures between states, and in some cases between different parts of a state. It will take some effort and work, but these challenges, and others as they arise, may be overcome and the LabCAT looks forward to becoming a full partner in a regional certification program.

Robert A. Lutz,[1] James B. Hewston,[1] David A. Savage[1] and Peter A. Spellerberg[1]

The AASHTO Accreditation Program: Serving the Hot Mix Asphalt Industry

---

Reference: Lutz, R. A., Hewston, J. B., Savage, D. A., and Spellerberg, P. A., "**The AASHTO Accreditation Program: Serving the Hot Mix Asphalt Industry,**" *Hot Mix Asphalt Construction: Certification and Accreditation Programs, ASTM STP 1378*, S. Shuler and J. S. Moulthrop, Eds., American Society for Testing and Materials, West Conshohocken, PA, 1999.

Abstract: The American Association of State Highway and Transportation Officials (AASHTO) has been accrediting hot mix asphalt testing laboratories under the AASHTO Accreditation Program since 1988. Laboratories must satisfy many quality system criteria, as well as participate in the AASHTO Materials Reference Laboratory (AMRL) on-site assessment and proficiency sample programs, in order to receive accreditation from AASHTO. Satisfying the requirements of the growing list of agencies specifying accreditation, often in various forms, made the first ten years a challenging experience for both the AAP and its accredited laboratories. AMRL and the testing laboratories involved in the AAP have dealt not only with multiple quality system standards but also with their evolving criteria as well. The development of the National Cooperation for Laboratory Accreditation will present more demands as the AAP moves into its second decade as an accrediting body. Analysis of AMRL proficiency sample data indicates that accreditation is beneficial to the construction industry and that the AAP meets its objective of promoting uniformity in the testing of construction materials.

Keywords: accreditation, asphalt, laboratory testing, quality

The American Association of State Highway and Transportation Officials (AASHTO) has been accrediting laboratories that test various construction materials, including hot mix asphalt (HMA), since 1988 [1]. As of the ten-year anniversary of the AASHTO Accreditation Program (AAP) in June 1998, there were nearly 400 laboratories accredited by AASHTO, including 292 laboratories that test HMA and related materials [2]. The growth of AAP has been partly due to an increase in the regulatory agencies that require the use of accredited laboratories in their QA/QC plans for HMA. This paper will focus on testing related to HMA and will discuss (1) the development of the AASHTO Accreditation Program, the criteria for AASHTO accreditation, and the current status of

---

[1] Assistant Program Supervisor, Assessor, Program Supervisor, and Assistant Manager, respectively, AASHTO Materials Reference Laboratory (AMRL), National Institute of Standards & Technology, 100 Bureau Drive, Stop 8622, Gaithersburg, MD 20899-8622.

AAP; (2) the impact of national and international quality standards on AAP and new technology related to asphalt performance; and (3) an evaluation of the impact of AASHTO accreditation on laboratory testing performance.

The AASHTO Accreditation Program (AAP)

The objective of the AASHTO Accreditation Program is to provide a mechanism for formally recognizing the competency of a laboratory to perform specific tests on specific construction materials. The accreditation granted by AASHTO is not a blanket recognition that applies to all services offered by a laboratory, but rather is an acknowledgment of a laboratory's demonstrated capability to perform specific tests and to satisfy accreditation criteria. The program for HMA uses the results of the on-site assessment and proficiency sample programs operated by the AASHTO Materials Reference Laboratory (AMRL) to judge a laboratory's ability to perform specific tests. AMRL was established in 1965, under the sponsorship of the AASHTO Highway Subcommittee on Materials (HSOM), to promote uniformity in testing in construction materials testing laboratories. AMRL is located at the National Institute of Standards and Technology (NIST) and operates under a memorandum of agreement between AASHTO and NIST. All accreditation decisions, however, are made solely by AASHTO through its HSOM.

Accreditation Criteria

There are four general criteria that must be met in order for a laboratory to become accredited by the AASHTO Accreditation Program. The laboratory must:

(1) meet specific personnel qualification requirements,
(2) be assessed by AMRL and correct any resulting deficiencies,
(3) test relevant AMRL proficiency samples and attempt to discover the reason(s) for test results beyond two standard deviations from the established mean value(s) and
(4) develop, implement and maintain a quality system that meets the requirements of AASHTO R 18, Standard Recommended Practice for Establishing and Implementing a Quality System for Construction Materials Testing Laboratories.

The laboratory manager must (1) be a full-time employee of the laboratory, (2) be a registered engineer, or a person with equivalent science-oriented education, or have experience having satisfactorily directed testing or inspection services and (3) have at least 3 years' experience in the inspection and testing of the materials. The supervising laboratory technician also must have at least 3 years experience in the inspection and testing of highway construction materials.

The laboratory must receive applicable AMRL on-site assessments and quality system evaluations at routine intervals. In addition, the laboratory must, within 90 days of the issuance of the formal assessment report noting any deficiencies, provide AMRL with satisfactory evidence that all deficiencies were corrected.

The laboratory must participate in all applicable AMRL proficiency sample programs (PSP). In addition, the laboratory must, within 60 days of issuance of the proficiency sample report, provide AMRL with a report summarizing the possible reasons for any poor results and the corrective action taken. Proficiency sample results beyond two standard deviations of the grand average values are considered to be poor results.

The laboratory must establish and implement a quality system that meets the requirements of AASHTO R 18. (A laboratory must satisfy additional criteria in order to be recognized by AASHTO for compliance with ASTM D 3666, Standard Specification for Minimum Requirements for Agencies Testing and Inspecting Bituminous Paving Materials, and/or ISO Guide 25, General Requirements for the Competence of Calibration and Testing Laboratories.)

The Role of the AASHTO Highway Subcommittee on Materials (HSOM)

AASHTO uses a management council approach in reaching decisions on accreditation as described in ASTM E 994, Standard Guide for Calibration and Testing Laboratory Accreditation Systems – General Requirements for Operation and Recognition. AASHTO has assigned responsibility for monitoring and administering the operation of AAP to its HSOM. AMRL acts as the technical advisor in compiling all necessary information resulting from the on-site assessment, quality system evaluation, proficiency testing, and communications from each laboratory that describe steps taken to correct identified deficiencies. Accreditation decisions are made by the Chair, AMRL Administrative Task Group (ATG) of the AASHTO HSOM, based on the information gathered by AMRL. Any appeals to those decisions are handled by the full ATG. AASHTO reviews a laboratory's accreditation status at three established times during the ongoing accreditation process: (1) prior to the issuance of the initial accreditation certificate, (2) every twelve months after the initial accreditation and (3) after each on-site assessment.

The Role of the AASHTO Materials Reference Laboratory (AMRL)

On-site assessment of testing laboratories is the most important function of the AMRL. The laboratory assessment program provides for the assessment of regularly participating laboratories at intervals of less than two years (usually 22 months), also known as a tour. AMRL laboratory assessors are supplied with a variety of equipment that includes vacuum gages, calipers, micrometers, timers, precision weights, thermometers and many miscellaneous items. The assessment of a laboratory consists of an observation of the test procedures, an examination of the apparatus used in performing selected physical tests and a review of a laboratory's quality system. Laboratory assessments are designed to accommodate AASHTO or ASTM test methods, or both.

The on-site assessment program was originally instituted to service the laboratories of the state Departments of Transportation, the Federal Highway Administration (FHWA) and other AASHTO sponsors. The program began to grow considerably in the 1980s when it was opened to any interested laboratory. By 1997, at the end of AMRL's twenty-first tour, participation had grown to 528 laboratories. That number is expected to surpass 630 during the twenty-second AMRL tour.

Distribution of proficiency test samples is the second most important function of the AMRL. AMRL has been distributing proficiency samples for more than thirty years. Proficiency samples are prepared for test methods that include, among others, viscosity graded asphalt, performance-graded binder, emulsified asphalt, HMA analysis, HMA design, HMA gyratory, fine aggregate and coarse aggregate. As with the laboratory assessment program, participation began to increase in the 1980s. Participation levels now include approximately 460 laboratories in the HMA programs, approximately 200 laboratories in the viscosity-graded and performance-graded binder programs, and more than 600 laboratories in the aggregate program.

The AMRL Proficiency Sample Programs (PSP) provide participating laboratories with the following benefits:

(1) A means of checking both instrument and operator performance under actual testing conditions.
(2) A means of comparing individual test results with the mean values of a large testing laboratory population. Corrective action may be taken when deviations from the mean occur.
(3) A means of evaluating the quality of a laboratory's test results, thereby reducing the risk of dispute due to testing errors.
(4) A means of documenting testing capability.

Current Status of the AASHTO Accreditation Program (AAP)

The AASHTO Accreditation Program has grown steadily since its inception in 1988. The participation status in the program, as of June 1998, by laboratory type and by material type, is as follows:

Number of accredited laboratories by laboratory type
- Independent/Commercial   248
- State/Federal   54
- Producer/Supplier   41
- Miscellaneous   25

Number of accredited laboratories by material type*
- Asphalt Cement   102
- Emulsified Asphalt   63
- Hot Mix Asphalt (HMA)   229
- HMA Aggregates   241
- Soil   226
- Portland Cement Concrete   246
- PCC Aggregate   258
- Hydraulic Cement   35

*Many laboratories are accredited for more than one material type.

Quality Standards

AASHTO R 18

At the heart of AAP is AASHTO R 18, which was incorporated in the AMRL Laboratory Assessment Program in 1993 and became mandatory in April 1994 for laboratories accredited through the AAP. AASHTO requires that accredited laboratories establish and maintain a quality system that satisfies R 18 requirements. These requirements include specifications for procedures and records associated with staff training and evaluation, and equipment calibration and verification. AASHTO R 18 also defines quality manual requirements, such as organizational charts, position descriptions, biographies and other documents. AASHTO R 18 is unique because it contains examples of the many prerequisites for standard operating procedures, equipment calibration and verification procedures, sample forms and many other items which assist laboratories in developing their quality systems.

ASTM D 3666

The current version of ASTM D 3666, which AASHTO in 1994 elected to include in the scope of the AAP, is similar to AASHTO R 18. AASHTO has encountered difficulties dealing with the current statement in the Personnel Qualifications section of ASTM D 3666 that requires laboratory personnel to possess an appropriate certificate from a national or state organization that meets the requirements of ASTM D 5506, Standard Practice for Organizations Engaged in the Certification of Personnel Testing and Inspecting Bituminous Paving Materials. AAP does not have the means to evaluate technician certification programs to determine whether they meet the requirements of ASTM D 5506. Additionally, at this time there are no organizations that recognize a certification program's compliance to D 5506. Therefore, AAP does not evaluate laboratories for compliance with ASTM D 3666 personnel requirements due to the subjective nature of these requirements.

AASHTO has noticed a recent surge in the number of laboratories seeking accreditation for ASTM D 3666. The AAP now includes 142 laboratories that have met the requirements of ASTM D 3666, or nearly fifty percent of all laboratories accredited by AASHTO for HMA, asphalt and/or HMA aggregate. The increase in the number of laboratories seeking recognition for compliance to D 3666 is largely due to requirements imposed by specifying agencies, such as the Federal Aviation Administration (FAA). The FAA's Standards for Specifying Construction of Airports specifies that laboratories meet the requirements of ASTM D 3666 for developing job mix formulas and the associated material acceptance testing [3].

ISO Guide 25

AASHTO has also been recognizing a laboratory's compliance to the requirements of ISO Guide 25. There are currently only five laboratories in the AAP that have ISO Guide 25 recognition. Compliance with ISO quality standards is essential if a laboratory wishes to have worldwide acceptance of its test results. In the future AASHTO anticipates that its list of ISO Guide 25 accredited laboratories will grow.

Impact Of National Activities on AAP

The Strategic Highway Research Program (SHRP)

About the same time as the AAP was established, SHRP was instituted to deliver major changes to the world of construction materials testing. For five and one-half years SHRP conducted a $50 million research venture in an effort to develop new and better ways to specify, test, and design asphalt materials. The development of national training centers to educate and train industry personnel in the correct use of these new performance-related equipment and test procedures soon followed. In 1993 AASHTO published its first set of provisional standards to quickly issue materials specifications and test methods resulting from the SHRP work. AASHTO provisional standards are standards that have been adopted by the AASHTO HSOM on an interim basis, not to exceed eight years. At any time during the eight-year period, the Subcommittee can ballot to convert a provisional standard to a full standard or can decide to discontinue the provisional standard. The eight-year period is used to refine the provisional standards based on comments received from the users and other reviewers.

During the early to mid 1990s AAP waited for the work from SHRP to unfold and the implementation phase to begin. Many producer and user agency personnel proceeded cautiously in the face of this dramatic change for the industry. The AMRL began implementing the SHRP technology in January 1996 by conducting laboratory assessments for four SHRP-related AASHTO provisional standards on an "informal" basis. The four tests included TP5, the dynamic shear rheometer (DSR); TP1, the bending beam rheometer (BBR); PP1, the pressurized aging vessel (PAV); and TP4, the gyratory compactor. The term "informal" meant that AMRL did not summarize findings in a formal, written report. As provisional standards, these test methods were considered works in progress and AASHTO elected not to add them to the scope of its accreditation program at that time. As the year passed it became obvious that these standards would retain their provisional status for at least a few more years. However, this did not prevent many agencies from specifying performance-graded binder. The publication of AASHTO PP 26, Standard Practice for Certifying Suppliers of Performance Graded Asphalt Binders, which specifies requirements and procedures for a certification system for suppliers of performance graded asphalt binder, heightened the interest for a formal assessment program. The demand for recognition of a laboratory's competency to perform these tests, through an accreditation program, increased by the end of 1996. AASHTO consequently added these four provisional standards to the scope of the accreditation program effective in January 1997. This decision committed the AMRL to change the SHRP assessments from informal to formal.

Judging a laboratory's competence to perform tests according to provisional standards is challenging. Though SHRP research began in 1987, scrutiny of the tests and the provisional standards continues even today. Each standard has undergone several revisions and more are anticipated. It is sometimes difficult for laboratories, and for

AMRL, to keep pace with these changes. The decision by AASHTO to offer accreditation for those four provisional standards, however, had several benefits.

The process of formal AMRL assessments provided an opportunity for many laboratories to learn about the details of these new methods and to express concerns about the new generation of testing equipment. Prior to this process, many laboratories were following the manufacturers' recommendations for equipment calibration and verification and testing protocols. In many cases, the recommended procedures did not meet the requirements of the provisional standards. The addition of the four provisional standards to the AAP also provided an opportunity for the laboratories to make comments about the standards; these remarks were compiled and were used in the revisions of the standards. Accreditation also quickly elevated the enrollment in the AMRL performance-graded binder proficiency sample program. As participation grew, sound estimates for precision were developed, based on the data from the many participants.

Future Considerations

Looking toward the future, the AASHTO Accreditation Program anticipates that increased emphasis will be placed on accreditation and ISO quality standards. Bituminous materials producers will seek ISO 9001, Quality Systems – Model for Quality Assurance in Design/Development, Production, Installation and Servicing, quality system registration and ISO Guide 25 accreditation of their QC laboratory in order to have their products accepted anywhere in the world. Proficiency sample providers, like AMRL, will be required to satisfy the requirements of ISO Guide 43, Development and Operation of Laboratory Proficiency Testing, in order for their sample programs to be acceptable to accrediting bodies. Accrediting bodies, like AAP, will be required to comply with the requirements of ISO Guide 58, Calibration and Testing Laboratory Accreditation Systems – General Requirements for Operation and Recognition, and impose the requirements of Guide 25 on the laboratories they accredit in order to be recognized by organizations like the National Cooperation for Laboratory Accreditation (NACLA) [4].

To address these challenges, AASHTO has incorporated an optional ISO Guide 25 quality system review within the scope of AAP. Several laboratories are currently recognized by AASHTO as complying with the requirements of Guide 25. In addition, AASHTO and AMRL are developing quality systems based on the requirements of ISO 9001 and ISO Guides 58 and 43 and plan to begin implementing the systems next year. AASHTO also plans to join NACLA when membership is opened and to consider seeking NACLA recognition in 1999.

As reliance on accreditation increases, AASHTO will modify the AAP and the programs of the AMRL to provide services that continue to meet the needs of specifiers and the asphalt industry.

Impact of AAP on Lab Performance

A laboratory must expend a significant amount of time and effort before it can become accredited by AASHTO. In addition to the demanding quality system requirements, laboratories must receive routine on-site assessments from AMRL and participate in the AMRL proficiency sample program (PSP), as previously described. Although it is often a struggle, many laboratories believe that the accreditation process improves their testing capabilities. In the search for an objective approach to evaluate the impact of the AAP, AMRL analyzed its proficiency sample data in an attempt to answer the following question: Do accredited laboratories demonstrate better testing performance than non-accredited labs?

To compare the performance of AASHTO accredited laboratories and non-accredited labs, AMRL proficiency sample data were separated into two groups: data from accredited laboratories and data from non-accredited laboratories. For each test, values were calculated for the mean, the standard deviation and the coefficient of variation of each of the two groups.

The data for each group were graphed in order to aid the comparison of the two groups. This initial visual examination, focused on the reproducibility of data, or between-laboratory variability, did not reveal any eye-catching distinctions. For HMA maximum specific gravity (AASHTO T 209 / ASTM D 2041), the testing performance of the accredited laboratories appeared to be equivalent to that of non-accredited labs (Figure 1). The majority of the data, for both groups, is concentrated and has a tight core.

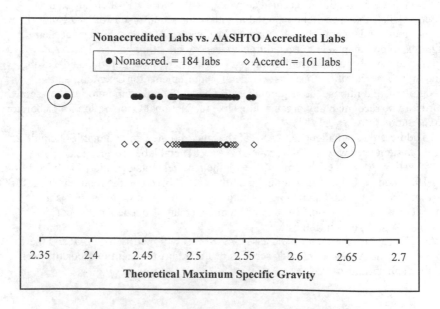

Figure 1 - *Comparison of Test Results from Laboratories Participating in AMRL PSP (HMA Sample Number 27)*

The same analysis was also applied to the following tests:

Hot Mix Asphalt (HMA) Tests
- Bulk Specific Gravity (AASHTO T 166 / ASTM D 2726)
- Marshall Stability and Flow (AASHTO T 245 / ASTM D 1559)
- Quantitative Extraction of Bitumen (AASHTO T 164 / ASTM D 2172)
- Gradation of Extracted Aggregate (AASHTO T 30 / ASTM D 5444)

Liquid Asphalt Tests
- Penetration (AASHTO T 49 / ASTM D 5)
- Specific Gravity (AASHTO T 228 / ASTM D 70)
- Kinematic Viscosity (AASHTO T 201 / ASTM D 2170)
- Viscosity at 60°C (AASHTO T 202 / ASTM D 2171)

Aggregate Tests
- Wash over 75-μm (No. 200) Sieve (AASHTO T 11 / ASTM C 117)
- Sieve Analysis (AASHTO T 27 / ASTM C 136)
- Specific Gravity of Fine Aggregate (AASHTO T 84 / ASTM C 128)
- Specific Gravity of Coarse Aggregate (AASHTO T 85 / ASTM C 127)
- L.A. Abrasion (AASHTO T 96 / ASTM C 131)

Again, this analysis did not reveal any significant difference between the performance of the accredited laboratories compared to the performance of the non-accredited laboratories. A few graphs seemed to indicate some improvement in testing performance for the accredited laboratories. A comparison of the variances, using an F test at a 95 percent confidence interval, revealed that any differences were not significant. Based on these results, the data did not support the expectation that accredited laboratories would produce a tighter group of test results. This was surprising because AASHTO accredited laboratories are subjected to recurrent on-site assessments from the AMRL, they must continually participate in AMRL proficiency sample programs, and they must implement a quality system.

Though the initial analysis of the data did not provide any distinct patterns, the analysis continued. The lone outlier in the graph of data from the accredited laboratories for maximum specific gravity of HMA (circled in Figure 1) opened the door to another approach. Further examination of the data revealed that this laboratory had been accredited by AASHTO in April 1997, just a short time before the HMA design samples were tested. This raised the question: Are the effects of the AASHTO Accreditation Program enhanced over time?

The data from the accredited group were then arranged into smaller subsets – laboratories accredited for more than one year, laboratories accredited for more than three years and laboratories accredited for more than five years – and again graphed so that performance could be visually evaluated. The delineation this time was clear and seemed to relate improved testing performance to the amount of time accredited by the AASHTO program. To test this theory, the process was repeated many times, with the other tests and other samples. The graphical analysis presented noticeable differences between the results from non-accredited laboratories and the results from laboratories accredited by AAP for five years or greater (Figures 2 through 7). In most cases – but not all – the data from those laboratories accredited for five years or more displayed less variance than the data from the non-accredited laboratories.

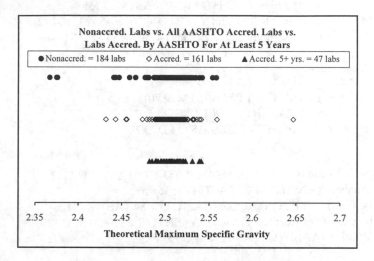

Figure 2 - *Comparison of Test Results from Laboratories Participating in AMRL PSP (HMA Sample Number 27)*

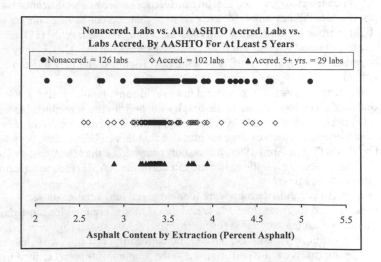

Figure 3 - *Comparison of Test Results from Laboratories Participating in AMRL PSP (HMA Sample Number 47)*

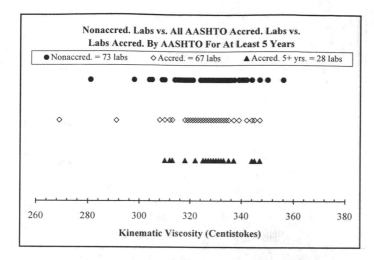

Figure 4 - *Comparison of Test Results from Laboratories Participating in AMRL PSP (Bituminous Sample Number 172)*

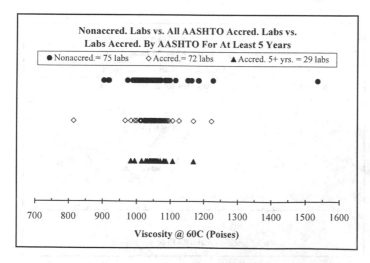

Figure 5 - *Comparison of Test Results from Laboratories Participating in AMRL PSP (Bituminous Sample Number 172)*

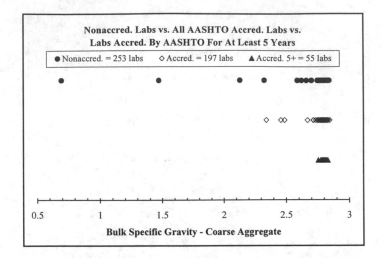

Figure 6 - *Comparison of Test Results from Laboratories Participating in AMRL PSP (Coarse Aggregate Sample Number 117)*

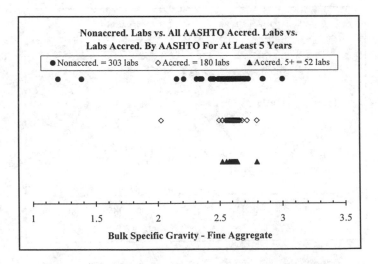

Figure 7 - *Comparison of Test Results from Laboratories Participating in AMRL PSP (Fine Aggregate Sample Number 120)*

In these instances, the coefficients of variation (1s%) also were lower for the group of laboratories accredited for five years or more (Table 1). F tests indicated that most of the differences were significant (Table 2).

Table 1 – *Reproducibility: Coefficients of Variation (1s%)*

Test	Non-accredited labs	Accredited labs	Labs accredited for 5 years or greater
HMA Max. Gravity (Sample 27)	0.81	0.79	0.49
HMA Max. Gravity (Sample 28)	0.75	0.78	0.51
HMA Asphalt Content (Sample 47)	10.80	8.96	5.92
HMA Asphalt Content (Sample 48)	13.92	8.44	7.53
HMA % Passing 12.5-mm Sieve (Sample 47)	7.20	1.58	1.52
HMA % Passing 12.5-mm Sieve (Sample 48)	7.87	1.89	1.11
Kinematic Viscosity (Sample 171)	6.58	3.52	2.72
Kinematic Viscosity (Sample 172)	13.17	3.59	2.77
Viscosity @ 60°C (Sample 171)	6.14	4.70	5.23
Viscosity @ 60°C (Sample 172)	7.46	4.43	3.22
Coarse Agg: Bulk Sp. Gr. (Sample 117)	5.96	1.76	0.41
Coarse Agg: Bulk Sp. Gr. (Sample 118)	6.09	1.63	0.42
Fine Agg: Bulk Sp. Gr. (Sample 119)	5.16	1.60	1.16
Fine Agg: Bulk Sp. Gr. (Sample 120)	4.82	2.01	1.28

Table 2 - *Comparison of Reproducibility Results from*
Non-accredited Labs and Labs Accredited for 5 Years or Greater

Test	F Test[1]	Differences Significant?[2]
HMA Max. Gravity (Sample 27)	0.43	No
HMA Max. Gravity (Sample 28)	0.94	No
HMA Asphalt Content (Sample 47)	2.45E-04	Yes
HMA Asphalt Content (Sample 48)	1.88E-04	Yes
HMA % Passing 12.5-mm Sieve (Sample 47)	2.95E-22	Yes
HMA % Passing 12.5-mm Sieve (Sample 48)	4.03E-30	Yes
Kinematic Viscosity (Sample 171)	2.9E-06	Yes
Kinematic Viscosity (Sample 172)	5.8E-13	Yes
Viscosity @ 60°C (Sample 171)	0.3195	No
Viscosity @ 60°C (Sample 172)	5.4E-06	Yes
Coarse Agg: Bulk Sp. Gr. (Sample 117)	2.0E-51	Yes
Coarse Agg: Bulk Sp. Gr. (Sample 118)	2.47E-51	Yes
Fine Agg: Bulk Sp. Gr. (Sample 119)	5.10E-23	Yes
Fine Agg: Bulk Sp. Gr. (Sample 120)	1.2E-19	Yes

[1]The one-tailed probability that the variances in the two groups are not significantly
different.
[2]Significance at a 95 percent confidence level.

   Although between-laboratory variability was the original focus of the study,
repeatability (within-laboratory variability or single-operator precision) was also
examined with the data from the AMRL proficiency sample program. After analyzing all
the data, several observations can be made.

(1) Laboratories that have participated in AAP for five or more years usually
    demonstrate improved between-laboratory testing precision (reproducibility)
    when compared to non-accredited laboratories.
(2) The same group of laboratories usually exhibits improved within-laboratory
    precision (repeatability) when compared to non-accredited laboratories.
(3) The impact of the AAP in improving test results does not seem to apply to all
    tests.

Why is the influence of AAP not seen on all tests? The graphical analysis for a few of the tests – Marshall stability and flow, for example – does not show the same pattern of time-related improved performance seen in the other tests. A possible explanation involves the inherently high variability for these tests. The maximum specific gravity test for HMA has a between-laboratory coefficient of variation (1s%) of less than one percent while the variations for the stability and flow tests are usually about twenty percent. It appears that accreditation has little effect on a test like the Marshall stability determination, where (d2s%) values approach sixty percent. Perhaps results from a test with poor precision cannot be significantly improved through the application of an accreditation program. Other factors may also mask the effects of AAP on testing performance. Many of the laboratories in the non-accredited group also participate in the AMRL programs, and some have developed a quality system. The only difference between some of the non-accredited laboratories and the accredited laboratories, therefore, is that the accredited laboratories have taken the step of seeking formal AAP recognition. In addition, a few of the laboratories in the non-accredited group were previously accredited laboratories. These points should not be lost when the results of this analysis are examined. It is difficult to evaluate the effects of AAP, and only AAP, when examining the data from the AMRL proficiency sample program.

Conclusions

During its first ten years, the AASHTO Accreditation Program has met the needs of its customers while maintaining a quality program. The requirements and the criteria of AAP were refined to shape the program as it grew; the program was also expanded to accommodate the needs of the industry. ASTM D 3666 was added to the scope of AAP so that laboratories could meet specifier requirements for conformance to that standard; ISO Guide 25 was added to the scope of AAP to include a quality system review that would meet international requirements; and four Provisional standards were added to the scope of AAP to accelerate the implementation phase of the innovative work resulting from SHRP. The next ten years, and beyond, will likely be no different in that AASHTO expects that the emphasis on accreditation, international quality standards and SHRP to continue.

A review of AMRL proficiency sample program data indicates that, in many cases, both repeatability and reproducibility improve for accredited laboratories over time. This suggests that improvement through participation in the AAP is a process, not an event. The effects of accreditation on the testing performance of laboratories are related to the precision level of each test. Differences between accredited laboratories and non-accredited laboratories are more apparent in tests with good precision. This distinction is not readily seen in tests with poor precision. Even the best laboratories cannot overcome the handicap of a test with inherently high variability. AASHTO accreditation was designed to be test specific and AMRL PSP data show that a laboratory accredited in a particular area (e.g. HMA) does not necessarily perform all tests well in that or other areas.

What does this analysis of AMRL proficiency sample data ultimately indicate? It reveals that the effects of the AAP are positive and sometimes dramatic, but they are not generally immediate. Accreditation is a process by which, over time, laboratories producing poor data can become good and good laboratories can become better. It is also not a means to transform a test with poor precision into a test with good precision. The AASHTO Accreditation Program is not a panacea for the HMA industry. It does, however, provide the foundation for improved performance and fewer testing discrepancies.

References

[1] Pielert, J. H. and Spellerberg, P. A., "AASHTO Materials Reference Laboratory – Thirty Years of Service to the Transportation Community," *TR News*, Transportation Research Board, Washington, D. C., March-April 1996, pp. 22-28.

[2] AASHTO, "AASHTO Directory of Accredited Laboratories," American Association of State Highway and Transportation Officials, Washington, D. C., June 1998.

[3] FAA, http://www.faa.gov/arp/part05.htm#p-401, Federal Aviation Administration.

[4] NACLA, http://ts.nist.gov/ts/htdocs/210/nacla/index.htm, National Cooperation for Laboratory Accreditation.

Donald W. Christensen,[1] Anne Stonex,[2] and Timothy Ramirez[3]

Experiences with Bituminous Paving Technician Training and Certification in Pennsylvania

Reference: Christensen, D. W., Stonex, A., and Raimirez, T., "Experiences with Bituminous Paving Technician Training and Certification in Pennsylvania," *Hot Mix Asphalt Construction: Certification and Accreditation Programs, ASTM STP 1378,* S. Shuler, and J. S. Moulthrop, Eds., American Society for Testing and Materials, West Conshohocken, PA, 1999.

Abstract: In 1995 the Pennsylvania Department of Transportation (PennDOT) decided to develop and implement a training and certification program for bituminous technicians. A committee of engineers from various organizations developed plans for the first series of courses, which were conducted in the winter of 1996. In the first year of training, seven field technician courses and three plant technician courses were presented. Related intensive laboratory courses were given on Superpave and Marshall mix design. Over 1,800 people had participated in the program as of March 1998; PennDOT, in conjunction with the Northeast Center of Excellence for Pavement Technology (NECEPT), is continuing to develop and refine this training and certification program to improve its effectiveness.

Keywords: training, certification, technicians, PennDOT

Introduction

The purpose of this paper is to present a summary of Pennsylvania's experience with bituminous paving technician training and certification. In 1995, the Code of Federal Regulations was modified to include a requirement that only "qualified" technicians be permitted to sample and test materials for purposes of acceptance testing or independent assurance on federally funded highway construction projects. This regulation has

---

[1]Assistant Professor of Civil Engineering, The Pennsylvania State University, Transportation Research Building, University Park, PA 16802.
[2]Research Engineer, The Pennsylvania Transportation Institute, Transportation Research Building, University Park, PA 16802.
[3]Senior Bituminous Materials Consultant, Materials and Testing Laboratory, Pennsylvania Department of Transportation, 118 State Street, Harrisburg, PA 17120.

39

generally been interpreted to mean that most paving technicians must be certified by the year 2000. Partly in order to meet this requirement, and partly to simply increase the quality of pavements in the state, PennDOT initiated a Bituminous Technician Training and Certification Program in 1995. Numerous problems were encountered in developing and implementing this program. This paper is meant to share the experiences of this program with other states and highway agencies that may be planning or in the process of engaging in technician training and certification. The authors hope that in so doing we can help others improve the effectiveness of their training and certification programs.

This paper includes a background section, in which the initial development and subsequent refinement of the Pennsylvania training and certification program are presented. This is essentially a chronology of events leading to the current program. The current Bituminous Paving Technician Training and Certification Program is then presented in some detail, including discussion of the formats for the various courses, the examinations, the facilities used, the instructors, and so forth. Based upon the authors' experiences, some general conclusions and recommendations are made concerning the training and certification of paving technicians.

Background

The Pennsylvania Department of Transportation has qualified field technicians for many years. PennDOT has used such qualified technicians in an accountable QC/QA and materials certification program. In addition, the use of a qualified plant technician assures PennDOT project personnel that the technician can perform the work according to specifications and established testing procedures.

Up to and through 1995, the eleven engineering districts within PennDOT individually qualified bituminous plant technicians. Bituminous field technicians were not required to be qualified. Typically, the plant technician was qualified by the district materials engineer in the district where the plant was delivering most of its hot mix. However, if a qualified plant technician crossed over a district boundary to do work, approval in that district was in some cases also required. Some districts accepted other districts' qualified technicians, others did not. The use of mainframe computer systems, where a list of qualified technicians could be stored and retrieved, improved the acceptance of qualified technicians across district boundaries.

The technician approval process consisted of the district materials engineer or his staff conducting an annual bituminous plant review and observing the plant technician at work. The review required the plant technician to demonstrate proficiency in performing normal or required testing, and in controlling the quality of the bituminous mixture, and the ability to work independently. PennDOT established a standard check-off list to be used for reviewing and observing a plant technician. Even with a standard review form, requirements and approval procedures varied from district to district. Candidate technicians were required to meet requirements of varying difficulty, depending upon the District performing the evaluation. These varying requirements and procedures sometimes resulted in contractor or producer frustration when technicians crossed district boundaries to work.

Federal Mandate for Training

In 1995, the Federal Highway Administration (FHWA) revised the Code of Federal Regulations concerning technicians involved with materials sampling and testing. The Code of Federal Regulations, 23 CFR, Part 637, Quality Assurance (QA) Procedures for Construction, issued June 29, 1995 contains the following requirement:

> "After June 29, 2000, all sampling and testing data to be used in the acceptance decision or the independent assurance program will be executed by qualified sampling and testing personnel."

In response to this new federal regulation, PennDOT's Asphalt Paving Quality Improvement Task Force (APQI-TF) discussed and decided to pursue certification as a method to qualify sampling and testing personnel. As background, the APQI-TF was originally established on October 24, 1994 to address bituminous paving issues that were directly related to quality. The APQI-TF consists of PennDOT and industry representatives with close cooperation from the FHWA. A working group established by the APQI-TF began development of PennDOT's Bituminous Technician Certification Program in the summer of 1995.

Initial Training Courses as Offered in 1995/96

The APQI-TF decided to have two primary courses: (1) a Field Technician Course; and (2) a Plant Technician Course. These courses would first be offered in the winter of 1995/96. The Field Technician Course would revolve more around construction technicians and sampling of bituminous materials on the job site. The Plant Technician Course would focus on plant operation to obtain the best quality hot-mix, sampling and testing within an asphalt concrete plant, and asphalt concrete mix design. However, the training working group felt that better quality flexible pavements could be produced if there was a better understanding among plant and field personnel concerning the nature of the other's work. Therefore, a short session on plant sampling and testing was presented in the Field Technician Course, and there was a session in the Plant Technician Course on construction of asphalt concrete pavements.

The Field Technician Course was largely based on a bituminous technician course developed by the joint AASHTO/FHWA/Industry Training Committee on Asphalt, and implemented by the National Highway Institute (NHI), the training branch of the FHWA. This course is entitled "Hot Mix Asphalt Construction" [1]. It is designed to be a 2-1/2 day course and is comprehensive in its treatment of asphalt pavement construction. The PennDOT training working group used many of the training materials developed for this course, but deleted some units and placed added emphasis on local practice. A variety of instructors was scheduled, mostly on a volunteer basis. Most of the instructors were PennDOT engineers, but a number of speakers from private corporations and paving industry groups were also used.

The Plant Technician Course was essentially developed from scratch, as no existing course meeting this need could be identified. NHI recently developed a Plant Technician Course, although it deals more with actual operation of the hot-mix plant, rather than the associated sampling, testing, and QC/QA procedures. The plant course discussed here was developed to conform to the same, NHI-type format as the field course, and included units on hot mix asphalt plants, asphalt concrete mix design, bituminous material tests and specifications, and other topics pertinent to bituminous paving technicians.

A variety of instructional materials were used for the plant course. Some, such as *Hot-Mix Asphalt Paving Handbook* [2], were also used in the Field Technician Certification Course. The Asphalt Institute's mix design manual, MS-2, was included in these course materials. Participants in both the field and plant courses were given a large, three-ring binder of current PennDOT specifications and test methods pertinent to asphalt concrete production and construction.

The field and plant courses were similar in general format. Both courses were delivered in hotel meeting rooms, with class enrollments of about 40. The courses were both 2-1/2 days long, ending with a three-hour examination. Examinations for both courses were open-book, and consisted of approximately 80 multiple-choice and true/false questions. The field examination was based in part upon a similar examination being given in Maryland for bituminous technician certification courses. Both examinations were ultimately compiled and reviewed by a number of PennDOT and industry engineers.

During the first training season, two laboratory courses were given. One was a 2-1/2 day Superpave Volumetric Workshop, which followed closely a short course jointly developed by FHWA and the Maryland State Highway Administration (SHA). The enrollment for these workshops is typically about 24 students, instructed in six groups of four students each. The other course was a 4-day, general laboratory practices course, which covered the fundamentals of aggregate and bituminous materials testing, and Marshall mix design. An introduction to Superpave was included in this course. The enrollment in this course was limited to 12 (six groups of two), to ensure as much hands-on activity as possible.

In general the working group planning and executing the initial series of training courses considered the program a success. No major problems were encountered in delivery of the courses. There was some delay in evaluating the examinations, because a decision was made to compile all examination scores before determining passing/failing grades, to ensure consistency and fairness to all course participants. The minimum passing score for both examinations was set at 70 percent. This resulted in approximately 90 percent of all participants passing each course. Failure rates were slightly higher for the Field Technician Course, but this was expected, as this was a more difficult examination. The Superpave Volumetrics Workshop was well received, as it had been when delivered in other states in the Northeast.

A number of problems was encountered during the first year of training activities. Using a number of volunteer instructors created logistical problems. Also, because of the large number of volunteer instructors, there was not always a clear focus on important information that would subsequently be included in the examination. There was anxiety among some course participants, many of whom had not been in a classroom environment for years. Some of the calculation methods presented during the class and/or included in the examinations were not fully consistent with PennDOT forms and methods. During compilation of the examinations, it was found that some of the questions were too difficult, or confusing, in some cases having more than one possible correct answer. A concerted effort was made to correct these various problems in the second year of the training and certification program.

Continued Development of Training Courses: 1996 through 1999

During the second year of training and certification courses as offered in 1996/97, a total of 16 courses were delivered: 7 Field Technician Certification Courses, 3 Plant Technician Certification Courses, 4 Superpave Volumetric Workshops, and 2 Laboratory Practices Courses. To improve consistency in instruction, and to ensure adequate coverage of all material included in the examinations, two instructors were hired to perform the bulk of the instruction. Exam questions were compiled and sorted by instructional unit, and made available to the instructors to make them more familiar with instructional objectives. An improved, custom set of slides was developed for the Plant Technician Course, covering many of the specifications and test methods covered. A detailed instructor's manual, similar to those used in NHI courses, was also compiled for this course. Some of the visual aids developed for the Plant Technician Course were also used on a limited basis in the Field Technician Course.

For both courses, a large number of short quizzes and workshops were designed and interspersed with other instructional activities. This served to better involve the course participants, and also to reinforce important ideas presented by the instructors. Additionally, these quizzes and workshops were designed to familiarize the participants with the format of the exam, to help prepare them and give them confidence. The examinations were slightly modified from the first year, by removing questions that were not clear or otherwise ineffective. Additionally, a decision was made to use computer grading, so that the examinations could be more quickly graded and the results made available to the participants. This provided an additional benefit in that statistics were provided on the overall difficulty and effectiveness of each examination question; this proved very valuable for reviewing and editing the examinations in preparation for the following training season.

The Superpave Volumetric Workshops were very popular, whereas the more general Laboratory Practices Courses did not have full enrollment. This clearly was because of the need for engineers and technicians to learn more about Superpave technology. Starting in the third year (1997/98) of training, the only laboratory course offered was an expanded version of the Superpave Volumetrics Workshop. This expanded course was based on the FHWA/Maryland workshop, but added 1-1/2 days to included expanded hands-on experience with fundamental laboratory techniques.

Continued refinements have been made in all courses for the third (1997/98) and fourth (1998/99) year of the PennDOT Bituminous Paving Technician Training and Certification Program. The examinations are reviewed every year, with the least effective questions replaced with new ones. Furthermore, a certain number of questions are replaced every year, to ensure that participants do not attempt to memorize examination questions. Visual aids are reviewed every year by PennDOT engineers and technicians to make certain they are up to date and accurate. The length of the training season has expanded considerably. During the first year, a total of nine courses was offered during the months of February and March. PennDOT and NECEPT are offering 31 bituminous training and certification courses for the 1998/99 calendar year, starting in November and running through March. Included for the first time in this training season are a number of refresher, or "update" courses. Technicians who have already been certified may regularly attend these 1-1/2 day courses to renew their certification. These courses in part review fundamental concepts, but also serve the important purpose of informing engineers and technicians of important changes in specifications and standard test methods.

At this point, the technician training program in Pennsylvania is now entering a "steady-state" mode, in which many technicians are now renewing certifications initially earned three years ago, with the first series of courses. The growth of the program, in terms of numbers and types of courses offered, is summarized in Table 1.

Table 1–*Development of PennDOT/NECEPT Bituminous Paving Training and Certification Program*

| | Number of Offerings by Course Type: | | | | | |
Year	Field	Plant	Laboratory	Superpave Volumetrics	Field Update	Plant Update
1995-96	5	3	1	1	---	---
1996-97	7	3	2	4	---	---
1997-98	9	3	---	8	---	---
1998-99	10	2	---	10	7	2

Description of Current Training Program

Currently, PennDOT recognizes four levels of bituminous technician certification:

- Bituminous Plant Technician in Training
- Bituminous Level 1 Plant Technician
- Bituminous Level 2 Plant Technician
- Bituminous Field Technician

The requirements for the various types of certifications involve various combinations of experience, training, and references by peers and supervisors. The specific requirements for each type of certification are summarized in Table 2.

These requirements are only a summary; other details apply, such as the time frame within which the required experience was gained. To maintain certification, technicians must continue to work a certain minimum number of hours in bituminous construction, and must also regularly attend various possible combinations of certification courses, update courses, and conferences.

Table 2–*PennDOT Bituminous Technician Certification Requirements*

Certification Type	Required Experience	Required Signatures	Required Training
Plant Technician in Training	None	• Level 2 Plant Technician • District Materials Engineer	None
Level 1 Plant Technician	1 year	• Level 2 Plant Technician • District Materials Engineer	*Plant Technician Review and Certification*
Level 2 Plant Technician	3 years and certification as Level 1 Plant Technician	• District Materials Engineer	Approved Volumetric Mix Design Course
Bituminous Field Technician	1 year	• Certified Field Technician or • Company Superintendent	*Bituminous Field Technician Review and Certification*

Format of Training and Certification Courses

As PennDOT and NECEPT enter the fourth year of the bituminous training and certification program, the three primary courses have now been refined by successive improvements after each year. These three primary courses are the Field Technician Review and Certification Course, the Plant Technician Review and Certification Course (Level I), and the Superpave Volumetric Mix Design/Expanded Laboratory Course. In addition, as mentioned previously, NECEPT has begun offering 1-1/2 day update courses for both field and plant technicians; these courses can be used towards periodic renewal of all certifications.

The current outline for the Plant Technician Review and Certification Course is shown in Table 3. A wide variety of instructional techniques and visual aids are used. Slides and accompanying notebooks developed by NECEPT provide coverage of various topics for which materials did not otherwise exist. The course also utilizes materials from outside sources, such as the Asphalt Institute's slides on mixture design; typically such materials are edited selectively to meet particular course objectives. Selected videos are used for units on construction, and on the operation of typical hot-mix plants. Several workshops, or exercises, are used to give participants an opportunity to perform various calculations, such as density and voids analyses. These workshops have been developed and refined to reflect correct procedures as accepted by experienced engineers and technicians in Pennsylvania. Furthermore, whenever possible, current PennDOT forms are used in performing such calculations. This makes many participants feel more at ease, and also serves to instruct course participants on the forms that should be used for various tests and calculations, and the proper way to complete these forms.

Table 3–*Outline for Plant Technician Review and Certification Course*

Day	Time	Topic
Day 1	8:00 - 8:30 AM	Check In
	8:30 - 8:45	Orientation and Introduction
	8:45 - 11:00	PennDOT Specifications and Test Methods
	11:00 - 12:00 noon	Overview of Mixture Design Process
	12:00 noon - 1:00 PM	Lunch
	1:00 - 2:00	Aggregate Tests and Gradation
	2:00 - 4:00	Aggregate Blending Workshop
	4:00 - 5:00	Specific Gravity and Absorption
Day 2	8:00 - 9:00 AM	Density and Voids Analysis
	9:00 - 11:00	PennDOT Bulletin 27
	11:00 - 12:00 noon	Superpave Mixture Design and Analysis
	12:00 noon - 1:00 PM	Lunch
	1:00 - 2:00	Quality Control/Process Control
	2:00 - 4:00	Hot Mix Asphalt Plants
	4:00 - 5:00	Introduction to Paving Operations
Day 3	8:00 - 8:45 AM	Review/Question and Answer
	8:45 - 11:45	Examination
	11:45	Dismissal

In addition to the extended workshops, numerous short, self-graded quizzes are used to reinforce important concepts, and to familiarize participants with the types of questions used on the examination. This also helps to keep the participants active and involved, and breaks up lengthy units into more digestible packets. The quizzes are especially useful as a complement to the several video portions of the training, where they are used to emphasize important information presented in the video that many students might otherwise forget.

Breaks are given as often as is possible, ideally every hour, though this is not always possible. Although some instructors might view such breaks as wasted time, they serve two important purposes. First, they allow the participants to better focus their attention during instruction. Second and more importantly, the authors have found that during these breaks, lively informal interactions often occur among course participants and instructors. These discussions often give participants a better understanding of various aspects of mix production, placement and compaction. Some participants compare alternate test methods, and discuss proper interpretation of various specifications with

PennDOT engineers and technicians. Such interactions are probably one of the greatest benefits of the certification program, in that they promote better communication among the various parties involved in flexible pavement construction. Without frequent breaks, such interactions would either not occur, or would be more formal and limited in nature, and less productive.

The Field Technician Review and Certification Course is very similar to the Plant Technician Course. The current outline for the Field Technician Course is given in Table 4. The overall structure of the course is similar to that for the plant course, and as mentioned earlier, is based upon the popular NHI course on bituminous pavement construction. Many of the materials are those used for this course, though not all of the units included in the NHI course are covered. PennDOT and NECEPT engineers have developed some special units for this course, such as that on PennDOT specifications and Pennsylvania Test Methods (PTMs). Some training materials are identical to those used in the plant course. Although the Field Technician Course does not involve as much quantitative work as the Plant Technician Course, there are several workshops. As with the plant course, frequent quizzes help to keep the participants involved and reinforce important concepts.

The Superpave Volumetrics/Expanded Lab Course is currently the only hands-on course offered as part of the PennDOT Paving Technician Training and Certification Program. As mentioned previously, there was poor enrollment for an earlier course emphasizing general mix design methods, including Marshall mix design techniques, and it has been dropped from the program. The popular Superpave workshop follows closely a common format, originally developed by the Maryland SHA and FHWA. It has, however, been expanded to a 3-day format, compared to the original 2-1/2 day format. This was done in order to incorporate more hands-on lab work, especially with routine procedures such as aggregate gradation and specific gravity. There is a large amount of emphasis on calculation of volumetric parameters, such as air voids, voids in the mineral aggregate (VMA), and voids filled with asphalt (VFA) in this course, which many PennDOT engineers feel is essential in performing mixture design and analysis. The current outline for this expanded Superpave Workshop is shown in Table 5.

Superpave workshops are approximately an even mixture of laboratory work and lecture. The classes are normally limited to an enrollment of 12, which is split up into six sets of partners. Each person is expected to perform all of the various procedures. It is essential to use at least two instructors, as often the class is split into two larger groups, each performing different activities. Sometimes guest instructors are used, such as for giving a brief discussion of the Superpave performance tests. This course is very popular; there is always a long waiting list to enroll in the Superpave workshops.

Examinations

Examinations for the Field Technician Course and Plant Technician Course are very similar. Both consist of 60 multiple-choice or true/false type questions. Three hours are allotted for completing the exams, and participants are free to use any and all course reference materials. Answers are entered on a computer coding form, and are grading by University Testing Services of Penn State. Computer grading offers many advantages for this type of certification exam. The turnaround time is quick, usually only a few days.

There are few mistakes in grading; the only errors are ones that occur in coding the exam keys. Such mistakes are quickly identified and easily corrected at the start of the training season. An additional benefit is that computer grading provides additional compilation of statistics on the difficulty and effectiveness of each question, and for the entire examination. University Testing Services provides the percentage of correct answers for each question, which is used to gage the difficulty of each question according to the following guidelines:
- 0 to 20 % correct, very difficult
- 21 to 60 % correct, difficult
- 61 to 90 % correct, moderately difficult
- 91 to 100 % correct, easy

The *biserial coefficient* is used to evaluate the effectiveness of each question. This parameter, which ranges from less than 0 to 1.00, is an indication of the degree of correlation between the response on a given question, and the overall test score. The following guidelines are suggested for evaluating question effectiveness using the biserial coefficient:
- negative values, ineffective
- 0.00 to 0.20, low effectiveness
- 0.21 to 0.40, medium effectiveness
- 0.41 to 1.00, high effectiveness

Table 4–*Outline for Field Technician Review and Certification Course*

Day	Time	Topic
Day 1	8:00 - 8:30 AM	Check In
	8:30 - 8:45	Orientation and Introduction
	8:45 - 12:00 noon	PennDOT Specifications and PTMs
	12:00 noon - 1:00 PM	Lunch
	1:00 - 2:30	Surface Preparation
	2:30 - 3:30	HMA Delivery
	3:30 - 5:00	HMA Placement
Day 2	8:00 - 10:00 AM	HMA Placement (continued)
	10:00 - 11:00	Joint Construction
	11:00 – 12:00 noon	Compaction
	12:00 noon - 1:00 PM	Lunch
	1:00 p.m. - 3:30	Compaction (continued)
	3:30 - 5:00	Troubleshooting
Day 3	8:00 - 8:30 AM	Review/Question and Answer
	8:30 - 11:30	Examination
	11:30	Dismissal

Table 5–*Outline for Superpave Volumetrics/Expanded Lab Course*

Time, Min.	Activity
	DAY 1, 1:00 TO 5:00 PM
45	Orientation and introduction to Superpave
45	Introduction to Superpave binder grading and Superpave implementation
30	Materials selection
15	Break
60	Materials selection (continued)
45	Gyratory compaction

DAY 2, 8:00 AM TO 12:00 NOON

Time, Min.	Activity
240	Superpave mix design workshop–FHWA Demonstration Project Workbook (includes two 15-minute breaks)
60	Lunch

DAY 2, 1:00 TO 5:00 PM

Time, Min.	Groups 1, 2, and 3	Groups 4, 5, and 6
45	Fine aggregate specific gravity and absorption	Coarse aggregate specific gravity and absorption
45	Coarse aggregate specific gravity and absorption	Fine aggregate specific gravity and absorption
30	Fine aggregate sieve analysis	Coarse aggregate sieve analysis
15	Break	
45	Coarse aggregate sieve analysis	Fine aggregate sieve analysis
60	Weight and batch aggregate for gyratory compaction specimens	

DAY 3, 8:00 AM TO 12:00 NOON

Time, Min.	Groups 1, 2, and 3	Groups 4, 5, and 6
30		Weigh specific gravity specimens; specific gravity and absorption calculations
30	Gyratory compaction	Rotational viscosity; overview and discussion of other Superpave binder tests
60		Resistance to moisture-induced damage (AASHTO T283)

(Table 5 continued)

Table 5–*Outline for Superpave Volumetrics/Expanded Lab Course* (continued)

Time, Min.	Activity	
	Groups 1, 2, and 3	Groups 4, 5, and 6
30	Weigh specific gravity specimens; specific gravity and absorption calculations	
30	Rotational viscosity; overview and discussion of other Superpave binder tests	Gyratory compaction
60	Resistance to moisture-induced damage (AASHTO T283)	
60	Lunch	

DAY 3, 1:00 TO 5:00 PM

Time, Min.	Groups 1, 2, and 3	Groups 4, 5, and 6
45	Fine aggregate angularity; coarse aggregate angularity	Sand equivalent demonstration; flat and elongated particles
45	Sand equivalent demonstration; flat and elongated particles	Fine aggregate angularity; coarse aggregate angularity
30	Bulk specific gravity of gyratory compacted specimens	Maximum specific gravity
15	Break	
30	Maximum specific gravity	Bulk specific gravity of gyratory compacted specimens
45	Complete testing and calculations	
30	Laboratory tour, including Superpave performance testing equipment	

DAY 4, 8:00 AM TO 12:00 NOON

Time, Min.	Activity
60	Density and voids analysis; review laboratory activities and mix design data
75	Superpave construction guidelines and issues, including QC/QA
15	Break
60	Course review and quiz
30	Quiz review; question and answer period; course evaluation

At the end of each training season, both exams are analyzed simultaneously for difficulty and effectiveness. Questions showing low effectiveness values are either thrown out or edited to improve their effectiveness. Usually, the reason for a particular question being ineffective is obvious. At the same time, some questions are arbitrarily removed and replaced, so that the exams are significantly modified and improved every year.

A third statistic tabulated for the entire exam is the Kuder-Richardson formula 20 value for reliability. This provides an overall reliability rating for the exam. Values should be over 0.80; the higher, the better. For the first year, reliability values were typically about 0.80. For the last year, reliability values ranged from about 0.83 to 0.88. If all exams were to be analyzed simultaneously, the overall reliability would be even higher.

For the 1997-98 training season, the average score for both the field and plant examinations was typically about 80 percent; the minimum passing score is 70 percent. About 91 percent of those taking the field technician examination passed, while slightly less–about 85 percent–of those taking the plant technician examination passed. Those failing the exam are permitted to retake the examination at a later date for a nominal fee. Additionally, if requested, a NECEPT representative will meet with the participant to discuss the results of the examination and suggest strategies for improved performance on the retest. For security purposes, exams are not returned to course participants, but are kept on file.

In general, there is satisfaction with the effectiveness of the current certification exams, both on the part of those administering the program, and the engineers and technicians who have completed the classes. The authors feel that the computer grading and related compilation of statistics is an important tool in delivering, developing and improving such certification exams. It is also important that all instructors are familiar with the examination for their course, so that they can focus their instruction on course objectives as evaluated on the examination.

Course Materials

As with the general layout, instructional and reference materials used vary depending upon the specific course. Instructors for the Field Technician Course make use of slides, videos, and the instructor's manual for NHI Course 13132, "Hot Mix Asphalt Construction" [1]. Participants in the Field Technician course are given the following materials:

- *Participant Manual* for National Highway Institute Course No. 13132: "Hot Mix Asphalt Construction" [3]
- *Hot-Mix Asphalt Paving Handbook* [2]
- Selections from PennDOT's *General Specifications: Publication 408M*
- Selections from PennDOT's *Field Test Manual: Publication 19*
- Handouts on various topics of special interest, such as specific gravity, metrication, applied statistics, and the various workshops given throughout the course

The instructors for the Plant Technician Course use slides and an instructor's manual prepared by NECEPT, with significant input and review by PennDOT personnel. For the section of the course on hot-mix plant operation, a video is used, called "Here's How a Hot Mix Plant Works" [4]. Also, selected slides and videos from the NHI course "Hot Mix Asphalt Construction" are used for a brief unit on asphalt concrete transport, placement, and compaction. Participants in the Plant Technician Course are given the following materials:

- A Plant Notebook compiled by NECEPT, which contains sections on PennDOT specifications and test methods, PennDOT mix design procedures, QC/QA plans and checklists; and some introductory notes on Superpave mix design
- *Hot-Mix Asphalt Paving Handbook* [2]
- Handouts to accompany slide presentations
- Handouts for various workshops

The course notebook is a very important aspect of the Plant Technician Review and Certification Course. This gives PennDOT, and also the Federal Highway Administration and other government agencies, an opportunity to disseminate new or revised specifications. It ensures that all course participants, and their associates at work, have a complete and up-to-date set of specifications and references. Reports, papers, and brochures of special interest can be included in the notebook, such as research reports on proposed construction or laboratory techniques. Some course participants have commented that these notebooks alone are worth the time and money invested in attending these courses.

For the Superpave Volumetrics/Expanded Lab Course, the instructors make use of Superpave slides developed by FHWA. These slides are used to present in introduction and overview to Superpave. The workshop itself is delivered using the FHWA's *Superpave Asphalt Mixture Design Workshop Workbook*, Version 4.2 [5]. In delivering this workshop, the instructor uses a set of overheads, which closely follows the workbook. Besides the Superpave workbook, course participants are also given the following materials:

- The Asphalt Institute's *Superpave Mix Design SP-2* [6]
- NAPA *Special Report 180 – Superpave Construction Guidelines* [7]
- A Superpave Notebook compiled by NECEPT, which includes the proposed revised N-design table from the Mixture ETG Fall 1998 meeting [8]; Selected portions from *AASHTO Provisional Standards* concerning Superpave; and several other papers and publications

As with the Plant Technician Course Notebook, the Superpave Notebook is a valuable tool for disseminating current information to engineers and technicians in Pennsylvania. The Notebook is updated yearly, with various documents of interest to paving engineers and technicians engaged in Superpave mixture design and analysis in Pennsylvania and neighboring states.

Facilities

During the first series of courses in early 1996, the courses were delivered in various moderately priced motels throughout the state, using conference rooms large enough to hold 40 to 60 people. The facilities of these hotels were sometimes a source of problems: inadequate audio-visual aids, noisy rooms, rooms that were too cold or too hot. These problems at times distracted participants. Holding the courses at various sites was convenient for some participants, since they did not have to travel far or stay overnight in a motel. On the other hand, this created logistical problems for the instructors, who were forced to spend many successive weeks traveling around the state. Besides the inconvenience, this schedule made compiling and preparing materials difficult. Making changes in overheads, slides, and handouts was nearly impossible, as office and staff support were generally not available.

Before starting the second year of courses, it was decided that most of the courses should be held at a single, central facility in State College, Pennsylvania. Although this had the disadvantage that many of the participants would have to travel several hours to attend the course, and possibly stay overnight at a hotel, there were a number of advantages:

- A site with good, consistent facilities would help maintain the quality of instruction;
- It would simplify planning and logistics;
- Instructors would be familiar with the facilities and the audio-visual aids; and
- Instructors would have office equipment and staff at their disposal for assistance in preparing, compiling, and editing course materials throughout the training period.

The facility selected was a large, new hotel and conference center in State College. This facility was specifically planned for technology transfer activities. It has 35 meeting rooms ranging in size from about 1,000 square feet to about 4,000 square feet, accommodating groups ranging in size from 10 to over 100. Each of the large meeting rooms was equipped with essentially identical sets of high-quality audio-visual equipment, including slide projectors, video projectors, and computer display projectors. These rooms had two screens for simultaneous projection of images from different video sources. All aspects of the audio-visual systems could be controlled electronically from the instructor's podium. The table and seating at these rooms were comfortable, and the rooms were quiet and comfortable.

The consensus of the instructors and the vast majority of course participants was that the superior facilities used in executing this second round of courses contributed greatly to the quality of instruction. Almost all participants supported continued use of these facilities, even though it meant some inconvenience and additional cost. The authors conclude that providing most training and certification courses at a high-quality, central location will provide much better results compared to holding classes over a wide area in a number of different facilities of varying quality.

NECEPT still provides training and certification courses at various locations throughout Pennsylvania, but the number of such courses is limited. These "travelling" courses are usually initiated at the request of local PennDOT engineers who perceive a large local demand for a specific training and certification course. Also, the main training program is given over the winter months when construction activity is light, enabling more people to attend. These travelling courses are usually given in the "off-season"–spring and summer–when there is more time for the instructors and support staff to plan the more difficult logistics of training at a remote location. This also gives instructors, some of whom are paid on a per-course basis, the opportunity to earn additional income over the entire year.

The Superpave Volumetrics/Expanded Lab Course is delivered at the Pennsylvania Transportation Institute (PTI), an interdisciplinary research facility at The Pennsylvania State University in University Park, PA. The lecture and workshop portions of this course are taught in PTI's conference room. The laboratory sections are taught in PTI's Pavements and Materials Laboratory. These facilities, originally developed for performing laboratory research, were upgraded several years ago for the purposes of delivering training courses. Several sections of the laboratory were designated as training areas, cleared of equipment that was not essential, and laid out specifically for training classes. Cosmetic improvements were made, such as painting and ordering new furniture. New equipment was ordered and set aside solely for training, in order to keep it clean and in good working order. Whenever possible, multiple sets of equipment were acquired, so that several groups could perform the same activity simultaneously. The main training area of the laboratory is roughly 900 square feet in area. This area seems comfortable for the typical class sizes taught in the Superpave laboratory classes.

Instructors

For the first series of courses, a wide range of volunteers from PennDOT, industry, and academia served as instructors. As a result, the quality of instruction varied. Some speakers did a good job, whereas others were difficult to hear, or were not well organized. A common problem was lack of congruence between the various instructors' presentations and the material on the examinations.

It was decided that for the subsequent training and certification courses, several instructors would be selected and paid typical rates for consulting services, according to their background and experience. The lead instructor was a well-known paving engineer, with experience in a wide range of agencies and good public speaking ability. One other professional engineer was selected to assist in the instruction. He was less experienced in paving, but had performed training in the past, and also had experience in transportation engineering and in working with PennDOT. Several guest speakers were scheduled to talk about specific topics, such as Superpave, and what was at that time the newly developed ignition oven method of determining asphalt content. A clear and well-documented course outline was developed, with objectives and example exam questions, to help all instructors and speakers stay focused on pertinent topics and skills.

Both instructors were offered opportunities to attend courses at other agencies, to improve their knowledge on certain topics, such as Superpave. They were also encouraged to attend laboratory courses at NECEPT, to improve their hands-on knowledge of various test procedures and related calculations. A third possible instructor was trained during this second year, by attending many of the courses, and occasionally assisting in the instruction. This eventually provided additional flexibility in scheduling courses, and also gave some insurance against emergencies that might prevent one of the other instructors from attending a course as scheduled.

In teaching laboratory courses, a different set of instructors is used. The lead instructor, providing lectures and overall course instruction, is a professional engineer experienced in mixture design and pavement construction. Laboratory instruction is however, provided by an experienced and highly qualified technician (NICET level III/IV). A technician from a local hot mix producer usually provides assistance. This helps put laboratory course participants at ease, and also ensures that instruction is consistent with local practice. The local technician and his employer enjoy the prestige this brings to their organization and the additional income during the winter months.

The authors suggest the following minimum qualifications for the lead instructor in classroom training and certification courses:

- At least 10 years' experience in the paving industry
- Significant, successful experience in education or public speaking
- Familiarity with local practice–state specifications, local climate and materials, and so forth
- Ability to devote substantial time in instruction on numerous courses, and in modifying and improving course during off-season

Assistant instructors should meet the following minimum qualifications:

- Five years' experience in the paving industry or a related field
- Some experience in education or public speaking
- Ability to devote time to teaching several courses
- Willingness to attend additional, outside training courses to improve knowledge

Lead laboratory instructors should have at least ten years' experience in bituminous materials testing, and should be NICET level III or IV certified, or equivalent. They should also be familiar with local test methods and specifications. Assistant laboratory instructors should have at least five years' experience in testing bituminous materials.

The authors have found that the effectiveness of the training and certification courses improves significantly as the experience of the instructors within the program increases. This is why all instructors should participate in as many courses as possible. In other words, a few instructors teaching a large number of courses will ultimately improve the quality of the courses, compared to using many instructors, each of which may only teach one or two courses a year. In the latter case, the instructors will likely never become fully familiar and confident with the course material and format. The instructors, especially the lead instructors, should be required to provide assistance in modifying the course materials during the off-season. These modifications typically involve improving visual aids, correcting errors, and keeping materials up to date with specification changes and new technology. As with the off-season travelling courses, this also provides instructors with activity and income during the spring and summer months. All instructors should be encouraged to attend other courses within the program that they are not necessarily involved with, and to attend one or two outside training courses. This will help improve the depth and breadth of their knowledge, and keep them up to date. This should also help improve their presentation and teaching skills, by observing instructors in other training and certification programs.

Instructor and Course Evaluations

Evaluations are given at the end for all courses, allowing participants to rate the quality of the course, the instructors, and the facilities. A short, one-page form is used.

Results are made available to the instructors, so that they are made aware of their respective strengths and weaknesses. Various PennDOT and industry engineers occasionally observe portions of courses to verify the quality and content of instruction. Providing an opportunity for participants to evaluate the course improves their perception of the program. A list of the questions used on NECEPT course evaluations is shown in Table 6.

Certification Panel

The Certification Board is an impartial panel created to handle certification program issues such as de-certification due to ethics problems or incompetence and appeals of certification status or program administrative policies. The makeup of the board is designed to keep any group from exerting undue influence by including two representatives from private industry, and one each from NECEPT, FHWA, and PennDOT. Protocols for disciplinary actions are in review, but the board has not yet had cause to meet.

Table 6–*Questions Used for Training and Certification Course Evaluations*

Please answer each of the following questions relating to the quality of this course, using the following rating scale:

1	2	3	4	5	6	7
very poor			OK			excellent

1. What was the overall quality of this course?

What was the overall quality of the instruction as provided by:
   2. Instructor 1:
   3. Instructor 2 (etc.):

5. What was the overall quality of the visual aids used during the course?

6. What was the overall quality of the manuals and handouts used during this course?

7. Rate the ability of the instructors to explain concepts and problems.

8. How relevant was the content of the course to your responsibilities and activities on the job?

9. What was the overall quality of the facilities (rooms, laboratories) used during this course?

PLEASE USE THE BACK OF THIS PAGE FOR SPECIFIC CRITICISM, COMMENTS, OR SUGGESTIONS.

Problems with the Current Certification Program

Although most people involved in the bituminous training and certification program feel it has been a success, there has been criticism of various aspects of the program. Such criticism has come from participants and also from PennDOT engineers who helped design the program. The two most serious criticisms were that the laboratory training and certification should involve rigorous hands-on proficiency testing, and that the written examinations should be more difficult. The use of proficiency testing was discussed during the early phases of the program, but was discarded because of the large resources needed to implement such a program. Such proficiency testing would ensure uniformly high quality in performing various bituminous materials test procedures. More difficult written tests would also help to ensure a higher level of knowledge among bituminous engineers and technicians. Such a difficult exam would, on the other hand, potentially reduce the pool of available engineers and technicians. Also, many experienced bituminous technicians have not been in a classroom environment for years; some will do poorly despite a reasonable mastery of subject matter simply because of poor test-taking skills and/or test anxiety.

Ultimately, the approach taken by PennDOT and NECEPT is a middle ground. The examinations are fairly rigorous, and typically 10 percent of participants do not pass the examination. Though the laboratory courses do not involve proficiency testing, all participants must perform a wide variety of test procedures and related calculations under close supervision of an experienced technician. It is possible that as the program evolves, the examinations will gradually become more difficult as paving technicians and engineers in Pennsylvania become more accustomed to the training and certification program. Similarly, laboratory proficiency testing could become feasible if more resources are made available for implementing the certification program.

Future of the Training and Certification Program in Pennsylvania

PennDOT's bituminous technician certification program has gained widespread acceptance throughout PennDOT, industry and the consulting field. The prestige and responsibilities for being a certified bituminous technician continue to grow. Much work was necessary to set up the program and improvements continue to be made. As each year passes, the bituminous technician certification program becomes less focused on how the program will logistically operate and more focused on the original objective of improving the quality of bituminous materials and paving. Some work is still needed to establish operating procedures, but it is anticipated that this effort will be completed within the next two years. PennDOT anticipates further emphasis in the future on incorporating into the training courses specification changes and timely issues of importance in the paving community.

It is anticipated that some regional acceptance and reciprocity of technician certification programs will occur among the states surrounding Pennsylvania. This may require additional revisions or modifications of PennDOT's Bituminous Technician Training and Certification Program. It may also require revisions to PennDOT's test methods, procedures, and/or specifications to make reciprocity practical. Although a monumental task, the potential rewards of such reciprocity in technician training and certification are great.

The Bituminous Technician Training and Certification Program has laid the groundwork for PennDOT's Aggregate Technician Certification Program. The operating logistics for this program were established quickly due to the success of the bituminous technician program. A similar training and certification program for portland cement concrete technicians is now under development.

Conclusions and Recommendations

The authors of this paper have been involved in various aspects of PennDOT's Bituminous Technician Training and Certification Program since its inception in 1995. The program is one of the largest of its kind, and is considered successful by the majority of PennDOT and industry personnel familiar with the program. Many feel that the quality of hot mix asphalt construction in Pennsylvania has already improved as a result of technician training and certification. The program is revised and updated every year, with the aim of constantly improving its quality and effectiveness. Based upon our experience, the following recommendations are made concerning bituminous training and certification programs:

- The NHI course "Hot Mix Asphalt Construction" is an effective basis for training and certifying field technicians;
- The Maryland/FHWA Superpave Workshop, with some additional laboratory work, is an effective laboratory course for bituminous technician training;
- The best possible facilities should be used for training and certification classes;
- Instructors should be well-paid, professional engineers experienced in paving construction, knowledgeable in local practices, and willing to devote significant long-term effort to technician training activities;
- Technician training and certification is a good forum for educating engineers and technicians on new technology, construction methods, and other topics of special interest to state and federal highway agencies;
- Training and certification programs should be updated and revised on a yearly basis, to constantly improve their effectiveness and keep them up to date; and
- Computer grading of examinations is a useful tool, providing quick turn-around of exam results, and valuable statistics for evaluating the difficulty and effectiveness of questions.

Acknowledgments

The authors would like to thank the many people who have made PennDOT's Bituminous Technician Training and Certification Program a success: instructors Carl Lubold, Pat Powers, and Perry Schram; Carlos Rosenberger of the Asphalt Institute; James Diversi, Tom Sikie, Bill Brookhart and Dick Kirk of PennDOT; Vince Angelo of Lehigh Paving, Paul Schrenk of Russel Industries; and many others who contributed their time and effort to this project.

References

[1] Joint AASHTO/FHWA/Industry Training Committee on Asphalt,"Hot Mix Asphalt Construction," NHI Course 13-132, American Association of State Highway and Transportation Officials and National Asphalt Pavement Association, 1993.

[2] US Army Corps of Engineer, *Hot Mix Asphalt Paving Handbook*, publication UN-13 (CEMP-ET), 1991.

[3] Joint AASHTO/FHWA/Industry Training Committee on Asphalt, *Hot Mix Asphalt Construction: Participant Manual,* American Association of State Highway and Transportation Officials and National Asphalt Pavement Association, 1993.

[4] Video University Productions, "Here's How a Hot Mix Plant Works," item No. 7, 3501 N. Happy Hollow Road, Independence, MO 64058.

[5] D'Angelo, J., Bukowski, J., and Harman, T., *Superpave Asphalt Mixture Design Workshop Workbook*, Version 4.2, Washington, D.C.: Federal Highway Administration, Office of Technology Applications.

[6] The Asphalt Institute, *Superpave Mix Design (SP-2)*, Lexington, KY: The Asphalt Institute, 1996, 117 pp.

[7] National Asphalt Pavement Association, *Superpave Construction Guidelines*, 5100 Forbes Blvd., Lanham, JD, NAPA Special Report 180, 1998.

[8] Bukowski, J., "Federal Highway Administration Mixture Expert Task Group Meeting Minutes," Fall 1998.

Kevin D. Hall[1] and L. Ray Pylant[2]

A First-Year Summary of the Arkansas Hot-Mix Asphalt Technician Certification Program

---

Reference: Hall, K. D., and Pylant, L. R., **"A First-Year Summary of the Arkansas Hot-Mix Asphalt Technician Certification Program,"** *Hot Mix Asphalt Construction: Certification and Accreditation Programs, ASTM STP 1378*, S. Shuler, and J. S. Moulthrop, Eds., American Society for Testing and Materials, West Conshohocken, PA, 1999.

Abstract: In response to Federal Aid requirements (23 CFR 637B, October 5, 1995) regarding certification of roadway technicians for quality control/quality assurance (QC/QA) testing, the Arkansas State Highway and Transportation Department (AHTD) sponsored the development of the Center for Training Transportation Professionals (CTTP) at the University of Arkansas, Fayetteville. Training and certification programs executed by CTTP include Hot-Mix Asphalt Field Technician (HMAC Tech) among others. Prerequisite to the HMAC Tech program is a training course in aggregates. The HMAC Tech program includes instruction and testing (written and performance) in the areas of sampling, gyratory compaction, volumetric analysis, asphalt content, and field density. As of November 1, 1998 eight HMAC Tech courses have been completed with a total of 150 persons attending, including 67 AHTD and 83 contractor employees. Of the 150 persons attending, 10 were unsuccessful, for a 93.3 percent pass rate. Student evaluations and anecdotal evidence indicate the program to be a resounding success.

Keywords: hot-mix asphalt, certification, quality control, quality assurance

---

[1] Associate Professor, Dept. of Civil Engineering, University of Arkansas, 4190 Bell Engineering Center, Fayetteville, Arkansas, 72701.
[2] Administrator, Center for Training Transportation Professionals, Dept. of Civil Engineering, University of Arkansas, 4190 Bell Engineering Center, Fayetteville, Arkansas, 72701.

Introduction

The Federal Highway Administration's Federal-Aid Policy Guide (23 CFR 637B) of October 5, 1995 established requirements that, by the year 2000, states receiving Federal Aid funds develop certification programs for ensuring roadway technicians performing quality control / quality assurance (QC/QA) activities are prepared to do so. In response, the Arkansas State Highway and Transportation Department (AHTD) sponsored the development of the Center for Training Transportation Professionals (CTTP) at the University of Arkansas, Fayetteville. The primary mission for the initial three-year CTTP agreement focuses on the development and delivery of technician-level certification programs in each of three major roadway construction activities – hot-mix asphalt concrete (HMAC), Portland cement concrete (PCC), and soils/earthwork. This paper describes program development and the initial eighteen months of operation, with particular emphasis on HMAC technician certification.

Certification Program Development

Discussions concerning the development of a certification program satisfying FHWA requirements were initiated by AHTD in late 1995. The Department of Civil Engineering at the University of Arkansas was invited to prepare a proposal for both the development of the program and its execution. An agreement was finalized and development efforts began in early 1996. By summer 1996 "pilot" programs in the areas of aggregates, soils, hot-mix asphalt, and Portland cement concrete were conducted at the University. The attendees of the pilot programs included AHTD engineering and field inspection personnel charged with the oversight of the certification effort. Training and certification curricula were revised following the pilot programs based on comments received by those attending; all training and examination materials were finalized in late 1996. The first training and certification programs for contractor and AHTD field testing personnel were conducted in February 1997.

Certification Issues

A number of key issues were identified and resolved during the development of the CTTP certification program for Arkansas. A listing and synopsis of some of these issues follows.

Certification Testing versus Training – the original concept of AHTD called for a certification testing program only. Through discussions with the University and an examination of similar programs in the U.S., it was decided to offer some level of training in addition (and prior) to certification testing to ensure that all attendees have full state-of-the-practice knowledge regarding particular testing methods. The training currently in place is designed to act as a "refresher" to attendees that currently work in a particular construction field, rather than as "primary" training in that area; this fact has caused some difficulty to attendees, as will be discussed later.

Certification Program Control / Authority – AHTD interprets Federal requirements regarding certification as applying not only to contractor technicians, but also to AHTD technicians and inspectors. Therefore, AHTD personnel must submit to the same training and testing as contractors (further discussion on this issue follows). To avoid any appearance of conflict-of-interest, total control of the certification program was given to the University of Arkansas in Fayetteville. The University (through CTTP, and in conjunction with AHTD in a strictly advisory role) determines training needs, curriculum, examination content, and allocation of program "seats" available to contractors and AHTD personnel. CTTP maintains all training and certification records. This level of autonomy apart from the state highway agency is viewed as one of the strengths of the Arkansas program. In this age of "partnership," contractors can be assured that all persons, regardless of affiliation, will be treated impartially.

Integration of Personnel – from the initiation of discussions concerning certification, the University of Arkansas insisted on having "integrated" training courses from the standpoint of contractor versus agency (AHTD). No course is offered that is strictly "AHTD" or strictly "contractor." This promotes at least two perceptions: all persons are trained and tested on identical material, and by agreeing on the method(s) for sampling and testing, potential conflicts in the field may be avoided.

Reciprocity – obviously, Arkansas is not the only state pursuing certification of technicians. One issue not yet fully resolved involves "accepting" the certification credentials issued by another state. Currently, decisions concerning reciprocity are made jointly by the CTTP Administrator and the QC/QA Technician Training program Director, on a case-by-case basis. Individuals seeking Arkansas certification based on credentials issued by another state provide documentation of curriculum included in their existing certification. Efforts are ongoing to establish some "baseline" curriculum with other states to serve as the basis for relatively automatic reciprocity.

Funding – because of the relative autonomy of the Arkansas program housed at the University of Arkansas, the issue of funding courses was critical. AHTD provided the initial funding for program development and necessary equipment purchases to host the training / certification courses. The ongoing agreement between CTTP and AHTD calls for a minimum number of courses to be offered per year (with a set number of seats reserved for AHTD personnel) at a fixed price per course to cover AHTD attendees. Contractor personnel attending certification programs are charged a fixed fee per person by CTTP, with no involvement by AHTD. During development, needs were forecast to allow a fee to be set for the entire initial three-year program contract without escalation.

Re-certification – testing methods for highway materials and construction do not see a large amount of change over short periods of time. However, some provision must be made to ensure persons remain state-of-the-practice in construction testing. The Arkansas certification in all programs is valid for four years, after which the person must submit to a re-certification process. The exact details of the re-certification process are not yet finalized.

Arkansas Hot-Mix Asphalt Technician Program

CTTP currently offers certification in three roadway construction areas: HMAC Technician, Soils/Aggregate Technician, and Portland Cement Concrete Technician. A fourth training course, Basic Aggregates, serves as a prerequisite to each of the certification areas (Figure 1). To obtain the Arkansas HMAC Tech certification, each person must successfully complete both the Basic Aggregates and HMAC Tech courses.

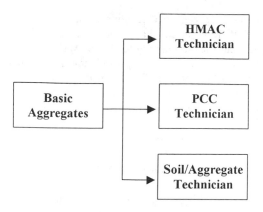

Figure 1 – *Arkansas Technician Certification Program Structure*

Basic Aggregates

The CTTP course in Basic Aggregates serves to provide attendees knowledge about aggregate sampling and testing to obtain properties used both directly in a QC/QA program and in subsequent testing and quality analysis processes for other materials, e.g. hot-mix asphalt concrete. Testing specifications included in Basic Aggregates represent those properties specifically listed in Arkansas' *Standard Specifications for Highway Construction* [1]. Table 1 lists the topics included in the Basic Aggregates course.

Table 1 – *Topics in Arkansas Basic Aggregates Course*

Topic / Subject	Applicable Specification[1]
Aggregate Field Sampling	AASHTO T-2
Reduction of Field Samples to Testing Size	AASHTO T-248
Crushed Particles in Aggregate	AHTD 305
Deleterious Materials	AHTD 302
Flat and Elongated Particles	AHTD 304
Moisture Content by Drying	AASHTO T-239
Washed Sieve Analysis	AASHTO T-11
Gradation / Sieve Analysis	AASHTO T-27
Specific Gravity – Fine Aggregate	AASHTO T-84
Specific Gravity – Coarse Aggregate	AASHTO T-85

The topics and testing methods included in Basic Aggregates are those aggregate-related properties that are relatively common to the three certification areas. Aggregate properties used specifically in a given certification area (i.e. fine aggregate angularity for hot-mix asphalt concrete) are reserved for that particular certification course.

Certification Testing – certification testing for Basic Aggregates combines a two-hour written exam with laboratory demonstration of proficiency in a number of testing methods. The written exam is "closed book" and consists of fifty multiple-choice format questions. The exam questions test not only familiarity with the specifics of testing specifications, but also basic computational abilities, such as determining aggregate gradation and specific gravity from test data.

The laboratory performance exam covers five primary tasks. These include:
- Splitting and quartering an aggregate field sample
- Determining the amount of minus-0.075 mm particles by washing
- Gradation / sieve analysis
- Determining the specific gravity of a fine aggregate
- Determining the specific gravity of a coarse aggregate

The minimum passing score on the written exam is seventy percent. On the performance exam, each student is allowed two attempts to successfully complete each of the five areas. These requirements are identical for each of the CTTP training and certification courses.

HMAC Technician

Curriculum – The testing specifications included in the HMAC Tech program represent those specifically listed in Arkansas' *Standard Specifications for Highway Construction* [1], supplemented by "special provision" specifications developed for Superpave-designed hot-mix asphalt concrete. The testing methods combine AASHTO standard methods with AHTD methods (Table 2).

Table 2 – *Testing Methods in Arkansas HMAC Tech Program*

Topic / Subject	Applicable Specification[1]
Sampling by Random Number Table	AHTD 465
Sampling Bituminous Paving Mixtures	AASHTO T-168
Sample Preparation – Superpave Gyratory Compactor	AASHTO TP-4
HMAC Bulk Specific Gravity	AASHTO T-166
HMAC Maximum Specific Gravity	AASHTO T-209
Percent Air Voids in Compacted HMAC	AASHTO T-269
Determination of VMA	AHTD 464
Binder Content using Nuclear Methods	AHTD 449/449a
In-Place Density using a Nuclear Gauge	AHTD 461
Moisture Damage in Compacted HMAC	AASHTO T-283

Conspicuously absent from the curriculum listing (Table 2) are Superpave consensus aggregate properties such as fine aggregate angularity and clay content, and Superpave source aggregate properties such as toughness and soundness. These tests may in fact be included in the HMAC Tech program in the future; however, the existing program focuses on day-to-day QC/QA testing requirements, rather than tests that are performed either prior to construction (during mix design for example) or at best infrequently during construction.

Another test being considered for inclusion into the HMAC Tech curriculum is the determination of binder content by ignition methods. Arkansas is moving closer to permitting the use of the ignition oven for QC purposes. When that procedure is added to construction specifications, the test method will be added to the certification course. It is noted also that the tests included in the Arkansas program focus on mixture properties and field density. No instruction or certification is currently offered for items such as pavement smoothness, asphalt plant inspection, etc.

Staffing – primary instruction for the HMAC Tech program is provided by faculty from the University of Arkansas, Department of Civil Engineering, supplemented by the CTTP Administrator. A single instructor is adequate under the current two-and-a-half day format; however, any increase in the time allotted for training would require additional instructional personnel. A drawback of using University faculty as primary instructors relates to availability. Depending on workloads, scheduling courses becomes difficult – when faculty are typically available (summer), construction personnel are hesitant to attend days of training. One bonus of hosting the course at a University is the "availability" of graduate Civil Engineering students for laboratory help in testing setup and cleanup.

Another staffing issue concerns the performance examination. As subsequently detailed, the HMAC Tech course includes five performance areas. For optimum efficiency in testing, each area is staffed with at least one testing Proctor. The number of testing "stations" is a function of the availability of Proctors and the amount of equipment available for use. Testing Proctors are each certified in the subject area(s) under which they serve.

Certification Testing – certification testing for the HMAC Tech program combines a two-hour written exam with laboratory demonstration of proficiency in a number of testing methods. The written exam is "closed book" and consists of sixty multiple-choice format questions. The exam questions test not only familiarity with the specifics of testing specifications, but also basic computational abilities, such as determining HMAC bulk and maximum specific gravity, calculating volumetric properties of mixes, and determining field density (percent compaction) from test data.

The laboratory performance exam covers five primary tasks. These include:

- Determination of bulk specific gravity of a compacted core
- Determination of maximum specific gravity of a loose mix
- Preparation of a compacted specimen using the Superpave gyratory compactor
- Preparation of a calibration specimen for the nuclear asphalt content gauge
- Determination of field density using the nuclear density gauge

The minimum passing score on the written exam is seventy percent. On the performance exam, each student is allowed two attempts to successfully complete each of the five areas.

The Arkansas Program's First Year

From the initiation of certification courses in February 1997 through October 1998, a total of eight HMAC Tech programs have been successfully completed. The total attendance at these courses was 150 persons, including 67 AHTD employees and 83 contractor personnel. A total of 38 different hot-mix asphalt materials, testing, and construction companies have been represented in the courses. In terms of pass/fail, 10 persons did not successfully complete all written and performance requirements to obtain the HMAC Tech certification, resulting in a 93.3 percent "pass rate." The unsuccessful candidates included both AHTD and contractor employees.

First-Year Observations

A combination of student course evaluations, an Annual Program Review (conducted in December 1997), and "follow-up" interviews with certified field personnel led to a number of observations concerning the relative impact and success of the program. A brief discussion of some of these observations follows.

Course Timing – the most frequent comment received by attendees of the HMAC Tech course involves the amount of material presented in the allotted time frame (approximately two-and-a-half days). Specifically, many students opined that too much material is included in the course. This issue is directly related to the experience of the person attending the course (discussed in the next section). Another complicating factor involved in this issue is the implementation of Superpave in Arkansas in 1997. Inclusion of Superpave technology (i.e. the gyratory compactor) forced even experienced personnel to "start over:" in a sense.

Prior Experience Level – one difficulty experienced in the first eight HMAC Tech courses was balancing the level of instruction to the level of field/lab experience of the attendees. While some of the attendees are very experienced in hot-mix related QC/QA testing, many were not. Indeed, an appreciable number of attendees confessed virtually "no" experience in hot-mix technology. As stated previously, the Arkansas courses were designed to be "refresher" training rather than primary training; however, both AHTD field offices and contractors send persons to the courses with inadequate prior preparation. To combat at least part of this problem, CTTP developed a Basic Math Skills self-evaluation to be distributed to all contractors and AHTD offices. This "test" allows an individual to gauge their preparation for the course in terms of fundamental calculations necessary for obtaining test results. Anecdotal evidence suggests that many individuals used the Basic Math Skills evaluation for review prior to attending the certification course(s).

Classroom versus Laboratory Training – certification involves demonstration of competence on a written exam and performance of laboratory tests. Accordingly, training involves both classroom and laboratory aspects. Striking a balance between the classroom and laboratory is an ongoing process. The Arkansas courses strive to provide as much laboratory "hands on" activity by attendees as is practical. This is particularly difficult with hot-mix technology; by necessity, the material to be tested must be kept hot and is not easily produced. In addition, limitations of laboratory space and equipment prevent all students from gaining all the experience they sometimes seek.

Course Attendees – to date, contractor personnel attending certification courses have represented the full spectrum of field personnel, from basic technicians to mix design engineers. However, AHTD focused on providing certification to the field personnel most directly involved with day-to-day QC/QA testing. As a result, very few AHTD field and resident engineers have received certification. Significant anecdotal evidence gathered from AHTD field personnel suggests that many resident engineers do not/will not support efforts by field inspectors to enforce proper sampling and testing techniques learned through and certified by the CTTP program. This issue must be resolved in the coming year.

Course Curriculum Content – hot-mix asphalt technology changes rapidly. Hot-mix asphalt certification must keep pace with that change. Two areas specifically identified for inclusion into existing HMAC Tech curriculum include a deeper understanding of the Superpave Gyratory Compactor (SGC) – calibration, for example – and alternate methods for determining the binder content of a mix, e.g. the ignition method. As discussed earlier, many students view the course as having too much content at present; inclusion of additional topics will force a lengthening of the time period allotted for training.

Overall Evaluation

The significant issues identified previously notwithstanding, attendees of the HMAC Tech training/certification course consistently rated the program "good" or "excellent" on course evaluation forms. Even experienced personnel consistently indicated the program provided value. Anecdotal evidence gathered from contractor and AHTD personnel indicates the certification of testing technicians has improved the quality of sampling and testing, and has improved the relationship between contractor and agency.

In terms of meeting demand, the eight courses completed to date were in fact all that were required for the time period February 1997 to November 1998. Indeed, the final two courses were conducted at less-than-capacity. However, a significant change takes effect January 1, 1999. Prior to that date, all persons involved with QC/QA testing were required to be certified, or "under the direct supervision of" a certified person [*1*]. As of January 1, 1999, all persons are required to possess a valid certification. Demand is expected to increase, making swift resolution of the issues raised in the first year's operation critical.

The Arkansas Program's Future

In the previous sections, a number of key issues were identified and discussed relative to the Arkansas HMAC Tech certification program. Four of those issues stand out as defining the direction of the program. A brief synopsis of each of the four follows.

Course Curriculum

When viewed in light of some other certification programs, the Arkansas program (in a sense) includes only the "bare necessities." The existing program includes those day-to-day tests specifically enumerated in the Arkansas standard construction specifications. Anticipated changes to the QC/QA system used in Arkansas (possibly accomplished in early 2000) may require additional tests or other QC/QA activities.

In terms of hot-mix asphalt technician certification, it is apparent that changes to the existing HMAC Tech program will force the course into a longer allotted time period. Inclusion of testing methods for Superpave, regarding both aggregates and mixes, necessitates a longer training effort prior to certification testing. In actuality, extending the HMAC Tech course to four days or more will place it on relatively even footing with many other States. The Arkansas course is among the "briefest" of many state's programs.

Reciprocity

This issue is closely related to the curriculum issue. As many HMAC contractors pursue work in multiple states, the expense of obtaining and maintaining multiple certifications seems unnecessary. State highway agencies in some regions of the U.S., such as the Northeast and Southeast, have initiated efforts to develop a "regional" certification, or at least identify a core curriculum set that could serve as a baseline for all states in the region. Using such a core curriculum could allow individual state's certification programs to offer programs aimed at only providing those few testing requirements and specifications unique to that state. Arkansas anticipates taking an active role in the possible development of a regional certification for the Southeastern U.S.

Personnel

As discussed earlier, anecdotal evidence suggests the Arkansas HMAC Tech program has not been successful in convincing many AHTD Resident Engineer level personnel of the value of the program. It is viewed as critical by CTTP that all AHTD Resident Engineers and Assistant Resident Engineers become certified under the program. The long term success of the certification effort requires that managers, as well as testing technicians, fully embrace the goals of the program – or at least support the concept of following a specification and "doing it right." Getting the field engineers into the program is a top priority in CTTP for the upcoming year(s).

Additional Certification Courses

The existing HMAC Tech certification program in Arkansas focuses on routine AC/QA testing specifications. Very little training (an no examination) is given relative to the *interpretation* of test results, and subsequent actions taken to correct problems. Many states offer advanced "levels" of certification for hot-mix construction, i.e. Quality Control Manager, etc. Arkansas has been successful in the basic technician level certification program to date. However, advanced levels of quality control certification will be necessary to ensure the highest quality asphalt concrete construction.

Summary

Arkansas certifies QC/QA technicians in hot-mix asphalt construction through a program administered by the Center for Training Transportation Professionals located at the University of Arkansas, Fayetteville. This program is sponsored by the Arkansas State Highway and Transportation Department, but is autonomous from AHTD in terms of operations and authority. Certification training and examination courses commenced in February 1997.

To date, eight HMAC Tech certification courses have been completed, with 150 persons attending. The first year-plus of operation led to the identification of key issues involved with QC/QA certification, many of which have been resolved successfully – but some of which remain. Those primary issues remaining include course curriculum (content), reciprocity with other programs, key personnel receiving certification, and extension of QC/QA courses and certification to advanced levels of quality control.

Student course evaluations and anecdotal evidence indicate the program to be highly successful thus far. Observations by CTTP personnel in field situations suggest certified individuals have improved sampling and testing skills. Based on comments received, the Arkansas State Highway and Transportation Department views the program as a valuable addition to the QC/QA process for construction.

References

[1] *Standard Specifications for Highway Construction*, Arkansas State Highway and Transportation Department, Little Rock, Arkansas, 1996.

Randy C. West[1] and Todd A. Lynn[1]

Certification and Accreditation Programs: A Contractor's Perspective

---

Reference: West, R. C., and Lynn, T. A., **"Certification and Accreditation Programs: A Contractor's Perspective,"** *Hot Mix Asphalt Construction: Certification and Accreditation Programs, ASTM STP 1378*, S. Shuler, and J. S. Moulthrop, Eds., American Society for Testing and Materials, West Conshohocken, PA, 1999.

Abstract: In recent years, the highway paving community has made notable strides to improve the quality of asphalt pavements that we all trust will ultimately result in longer lasting roads. The most conspicuous of these improvements is Superpave which has ushered in better asphalt binder characterization and mix design procedures. Coupled with Quality Control/Quality Assurance (QC/QA) programs it has also created a need for education and reeducation that will likely have as much of a positive impact on the HMA industry as the new test methods. All of these developments make technician certification and laboratory accreditation programs indispensable. This paper provides a contractor's perspective of certification and accreditation. The need for standardized certification and accreditation programs is presented and discussed.

Keywords: certification, accreditation, QC/QA, Superpave, pavement performance

Background

Several changes in the HMA industry have created an enormous demand for training and certification of technical personnel and accreditation of laboratory facilities involved in materials testing. The most conspicuous change occurring in the industry is the implementation of the Superpave system. One of the opportunities of Superpave implementation has been standardization of methods, terminology, and specifications. However, many of the new test methods and specification requirements continue to evolve as the industry gains experience and as research continues. Also, state highway agencies have implemented the Superpave system at different times and at different rates.

Another process undergoing continuous evolution is the development of specifications regarding field control and acceptance of construction materials. Keeping up with the implementation or revision of Quality Control/Quality Assurance (QC/QA) programs is a daunting task for agencies and contractors. New policies in the federal regulations on Quality Assurance Procedures for Construction (23 CFR Part 637) have been the catalyst for a new round of changes. The amendment to this regulation in 1995 [1] added the flexibility of using contractor test data in acceptance decisions. In the

---

[1] Director and materials engineer, respectively, APAC, Inc., Materials Services, 3005 Port Cobb Drive, Smyrna, GA 30080.

present trend of smaller and more efficient governments, many state highway agencies have embraced this policy change as an opportunity to reduce staffing levels of agency personnel. More responsibility has been placed on contractors to assure that quality materials and construction practices are utilized in road building.

Additionally, a recent directive by the Federal Aviation Administration [2] emphasized enforcement of the qualification of laboratories used for design and acceptance testing of HMA on federally funded airport construction projects. The directive requires that labs used for mix design and acceptance testing be certified to meet ASTM Standard Specification for Minimum Requirements for Agencies Testing and Inspecting Bituminous Paving Materials (D 3666).

Each of these changes has increased the need for training of materials technicians and engineers. The FHWA, state agencies, industry groups, and academic institutions have offered a plethora of short courses, seminars, and mix design classes on Superpave. In addition, state agencies that are implementing or have existing QC/QA programs also utilize training courses in their certification requirements for technicians.

Recent interest in regional qualification programs for personnel and testing labs has begun to result in action. Six states in the northeast have joined together in an effort to develop a regional certification program known as the New England Transportation Technician Certification Program. Also, thirteen western states, known as the Western Alliance, have initiated discussions toward a similar goal. These regional certification groups may provide a much more effective means of administering these programs due to sharing of program costs, utilizing a greater pool of expertise for instruction, and through sharing of solutions. Many contractors will also benefit from regional certification programs since technicians who work in multiple jurisdictions will not have to become certified by each agency.

Purpose of Certification and Accreditation Programs

Occasionally it is worthwhile to question why certain requirements exist. It is equally important to question and evaluate current programs designed to fulfill those requirements. Regarding both technician certification and laboratory accreditation, the purpose of the requirements is basically to assure the proper execution of the tests that are used to judge the quality, acceptability, and payment of construction materials. A reality of this industry is that the accuracy of test results used for these purposes is largely dependent on the skills of the technician and the equipment used to conduct the test. Therefore, qualification programs are needed to minimize the effects of technicians and equipment on test results.

From the perspective of a contractor, technician certification and laboratory accreditation programs are beneficial for several reasons. First, these programs ensure a fair set of standards for judging the competency of our personnel, our competitors' personnel, and agency or consultant personnel. Second, the programs establish uniform methods and equipment so that potential differences are minimized or eliminated. Third, and most important, it is essential that QC testing be as accurate as possible so that the appropriate reactions to changes in materials or production operations can be made.

In most states, the development of technician certification programs has coincided with the implementation of quality control/quality assurance specifications. Some states began using QC/QA specifications, and thus began their certification programs, over twenty years ago. Other states have only recently begun implementation of QC/QA specifications and certification programs. As a result of the evolving specifications and the time differences between implementation schedules of the states, the certification programs have developed independently.

The revisions made to the FHWA policy on highway construction in 1995 (23 CFR Part 637) [1] are having a significant impact on this issue. The revisions give the state highway agencies the option of using contractors' quality control testing as the acceptance testing provided that such testing shall be performed in "qualified" laboratories by "qualified" personnel. The terms certification and accreditation, in reference to personnel and facilities respectively, were deleted from the draft document and replaced with the term "qualified". Nevertheless, the intent of the ruling is "... to provide adequate assurance that the public is receiving the desired quality in the product produced by the contractor." [1]. Unfortunately, the ruling does not explicitly describe a qualification method. Instead, this responsibility is given to the individual states, which then have considerable latitude in administering their own qualification programs. As a result, technician certification and laboratory accreditation programs are likely to continue to develop independently in many states.

Standardization of Certification and Accreditation Programs

"Qualified" Personnel

Although the programs for qualifying technicians have been developed and exist independently, most programs are very similar with respect to key elements and levels of certification. For example, most programs have a field technician qualification level for QC and QA testing, a roadway technician qualification level, and a mix designer qualification level.

The elements or steps necessary to achieve these levels of certification typically include documentation of experience, participation in training classes or workshops, a written exam, and a hands-on performance exam. Certification programs also usually include policies on appeals, recertification, and decertification.

Since many of the technician certification programs are similar, it is plausible to have a system in which a standardized certification is valid in any state. As with a driver's license, individuals would qualify and take the required exams in his/her state of residency, and would be able to use that certification in any other state. Such systems already exist in the concrete industry with the American Concrete Institute (ACI) field and laboratory technician certification programs [3,4]. Many states currently recognize or require the ACI technician certification for contractors performing concrete work.

A standardized national or regional HMA technician certification program would benefit agencies and the industry. Much of the effort in running a certification program is administrative type work. This includes scheduling and arranging training classes, grading exams, issuing certificates, and management of records. These responsibilities are easily handled by an outside agency, institution, or association. Several state departments of transportation which currently take this approach have found that it removes a significant administrative burden from the state's central technical staff, giving them more time to address important technical issues. Consolidation of program administration services among many states would further improve efficiency and significantly reduce administrative costs.

The trend of consolidation continues to spread through virtually all industries and businesses and it is certainly true of the hot mix asphalt construction industry. A significant number of contractors now have operations and/or pursue contracts in more than one state. APAC currently has divisions operating in fifteen states. Although APAC is larger than the average contractor, its divisions operate much like many smaller local-market contractors. About one third of the APAC divisions do business in more than one state. For these divisions and many other similar contractors, dealing with multiple certification, accreditation or qualification programs has been accepted as a reality of doing business. However, the duplication of requirements for obtaining certifications in adjoining states is an unnecessary waste of time and money.

The single largest hurdle to a standardized technician certification program is differing sampling and testing methods among the states. Despite the existence of the national ASTM and AASHTO standards, most states have a separate set of methods and sometimes specialized equipment used only in their particular state. Reasons for states having different methods vary; some reasons are well justified and some reasons are questionable. With the implementation of the Superpave system, our industry has a great opportunity to standardize procedures, specifications and terminology. This is the chance to start with a new set of procedures that can be taught and learned by everyone. A standardized qualification program could first be initiated involving basic QC practices and tests (Table 1). The program could then be expanded to include qualification of roadway and mix design personnel. While it is important to allow the Superpave system to evolve through research and experience, states must be careful not to independently alter the procedures. Common practices and terminology will facilitate sharing of knowledge and experiences, which will produce incremental refinements that will ultimately improve pavement performance.

Laboratory Qualification

Proper facilities and equipment are as important as qualified personnel. Presently, laboratory qualification requirements vary significantly among agencies. Some states do not have formal qualification requirements for labs whereas other states or agencies, like the Arizona Department of Transportation, require labs to be accredited by AASHTO [5]. Significant elements that must be satisfied for the AASHTO Accreditation Program (AAP) are:

1. Inspection by the AASHTO Materials Reference Laboratory (AMRL)
2. Quality System Documentation Manual
3. Participation in AMRL Proficiency Sample Testing Program

Table 1 - *Basic Practices and Tests*

Practice or Test	ASTM Method	AASHTO Method
Sampling Aggregates	D 75	T 2
Reducing Aggregate Samples to Test Size	C 702	T 248
Sampling Asphalt	D 140	T 40
Sampling Hot Mix	D 979	T 168
Moisture Content	C 566	T 255
Aggregate Washing	C 117	T 11
Sieve Analysis	C 136	T 27
Superpave Gyratory Compaction	-----	TP 4
Bulk Specific Gravity	D 2726	T 166
Rice Specific Gravity	D 2041	T 209
Air Voids Calculation	D 3203	T 269

Under D 3666, laboratory qualification is very similar to the AAP but more specific personnel requirements are placed on management, the laboratory supervisor and technicians.

Most individuals that have experience with the AAP or D 3666 would agree that they are very worthwhile processes; however, some of the requirements are not practical for field labs. Nonetheless, field labs and state district labs are just as important as central design labs and should not be overlooked in the laboratory qualification program. All HMA is produced and adjusted, and projects are constructed, on the basis of results from field and district labs.

The federal policy provides several necessary requirements for qualification of laboratories used in acceptance testing (i.e. field labs). All such laboratories are to be periodically reviewed under an Independent Assurance (IA) program which shall include sampling and testing observations, equipment calibration checks, and split sample testing comparisons or proficiency samples. In cases where contractor quality control testing is used for acceptance, this requirement also applies to the contractor's lab. This policy allows the states to simplify the qualification requirement yet still maintain an appropriate system of checks on the field labs. However, this approach does not promote standardization. As with technician certification, a standardized accreditation program for field laboratories will not be possible if states are permitted to develop their IA programs independently.

Summary and Conclusion

As a result of changes to 23 CFR Part 637, "qualification" of testing personnel and laboratories will be required. Superpave implementation, coupled with QC/QA programs in some states, has created a need for education and reeducation that will likely have as much of a positive impact on the HMA industry as the new test methods. These developments make technician certification and laboratory accreditation programs indispensable. Standardized testing procedures and qualification programs would certainly foster better communication, information dissemination, and improved methods and techniques. However, the CFR ruling allows for considerable latitude concerning the administration of training and inspection programs intended to comply with the regulation. Consequently, it is feared that the result will be a proliferation of programs nationwide as is the case with existing test procedures. This has already become evident as many states have initiated or continue to develop qualification programs independently. Undoubtedly, this will place a burden on contractors and consultants who conduct business for more than one agency. More importantly, we may miss a great opportunity to progress and advance the status of the HMA industry.

References

[1] Federal Register, 23 CFR Part 637, Vol. 60, No. 125, 29 June 1995.

[2] Federal Aviation Administration, Advisory Circular 150/5370-10A, Part V, 8 September 1997.

[3] American Concrete Institute, *Technician Workbook, ACI Certification Program for Concrete Field Testing Technician – Grade I*, ACI International, Fourteenth Edition, 1996.

[4] American Concrete Institute, *Technician Workbook, ACI Certification Program for Concrete Laboratory Testing Technician – Grade I and Grade II*, ACI International, Fourth Edition, 1997.

[5] Arizona Department of Transportation, Highways Division, Materials Group, Quality Assurance Section, *System for the Evaluation of Testing Laboratories,* Arizona Department of Transportation, 8 April 1998.

Ahmed Farouki,[1] Michael A. Clark,[2] and John D. Antrim[3]

Basic Elements in the Design of a Certification Program for Hot-Mix Asphalt Construction Personnel

---

Reference: Farouki, A., Clark, M. A., Antrim, J. D., **"Basic Elements in the Design of a Certification Program for Hot-Mix Asphalt Construction Personnel,"** *Hot Mix Asphalt Construction: Certification and Accreditation Programs, ASTM STP 1378,* S. Sculer, and J. S. Moulthrop, Eds., American Society for Testing and Materials, West Conshohocken, PA, 1999.

Abstract: Competent personnel along with up-to-date practices are essential for quality engineering, construction and maintenance services. Deciding on whether an individual has the knowledge, skill and ability needed to satisfactorily perform identified tasks can be done via a multitude of processes which offer varying degrees of reliability and discrimination. A convenient and highly reliable method to determine the individual's competence is acceptance of a third-party evaluation of the individual's knowledge and skills.

One such third-party evaluation is the job-task competency-based engineering technician certification model which was developed by the National Institute for Certification in Engineering Technologies (NICET) during 1976-1979 in fulfillment of a contract with the FHWA to provide "a nationwide system for enrolling, testing and certifying technicians in engineering activities related to transportation".

The foundation of the NICET certification model is a "Practice Analysis." A technical advisory panel representing the industry concerned identifies and develops a NICET-certified technician profile, typical job assignments and responsibilities, and specific job tasks along with the respective time frame during which the technician should become proficient in these tasks. An industry wide validation is then conducted which refines scope and content of the certification program. Exam questions are written by practitioners to reflect on-the-job skills and knowledge of individuals from the entry level (or trainee) through the more advanced (senior) and supervisory levels. Finally, a pilot test is conducted using a representative group from the potential program users to determine the appropriateness of the exam questions.

The requirements for NICET certification are somewhat unique in that more is required than just meeting an examination requirement. Equal weight is also placed on satisfying requirements for relevant work history and for verification of actual on-the-job performance by a supervisor. This model has been in use nationwide for the past twenty years and has been proven to be an effective, reliable, fair, flexible and efficient tool for evaluating the competency of engineering technicians.

---

[1]Senior Civil Engineer, NICET, Alexandria, VA 22314.
[2]Deputy General Manager, NICET, Alexandria, VA 22314.
[3]General Manager, NICET, Alexandria, VA 22314.

Keywords: certification, competence, examination, technician, quality assurance, NICET

Introduction

Quality assurance and quality control cannot exist without properly trained, competent personnel determining compliance with project plans and specifications. Competence is defined as a combination of skill, knowledge, and ability that allows one to perform tasks properly and efficiently. Those engaged in testing and inspection must have:

- an understanding of engineering principals and materials science through appropriate combinations of education and training;
- sufficient experience in materials and process acceptance sampling and testing to be considered capable of dealing with a variety of construction materials under varying conditions;
- sufficient experience with construction processes to objectively observe and report conformance with plans and specifications.

Defining the qualifications of testing and inspection personnel is a necessary step, but it is not the only step. There needs to be in place an evaluation process that determines whether or not the inspector or tester has the skills, knowledge and abilities needed to satisfactorily perform his or her duties. Evaluation processes can range from ones that involve minimal effort and reliability to others that involve comprehensive efforts with varying degrees of discrimination.

A convenient and reliable method for stakeholders to determine the competency of an inspector or tester is the acceptance of a third-party evaluation of the technician's knowledge, skills and abilities (KSA). NICET offers such an evaluation by way of its certification model. The model requires satisfying a written exam requirement, verification of actual task performance by the applicant's direct supervisor, minimum relevant work experience and a general character and performance reference in the form of a personal recommendation.

The NICET Model

NICET, through its job-task-competency certification model, certifies engineering technicians who have acquired expertise and job responsibility in a technical specialty area. Significant features of the model include:

- Criteria that puts equal weight on satisfying a relevant work experience requirement, an applicable written examination requirement and a requirement for verification of on-the-job performance by a supervisor.
- Four levels of certification which recognize a career path of upward progression in expertise and responsibility of the individual.
- A program structure that uses stand-alone modules (called "work elements") to assess the candidate's KSAs. Each work element describes a relevant knowledge unit or an on-the-job process. The work elements are assigned to one of eight categories that classify job tasks from entry level to senior level and as common or specialized tasks.
- A program structure which allows the applicant to select, with considerable freedom, the work elements which will appear on their examination. Other than a limited

number of mandatory work elements, there are more work elements available in each of the eight categories than are needed to satisfy a particular certification examination requirement.
- A program that does not restrict who may sit for an examination.

Program Development

The development of the scope and content depth of a NICET certification program for a particular technical specialty involves several sequential and interdependent components as follows:

The Technical Advisory Committee

Once the need for a certification program is identified, a technical advisory committee representing the stakeholders is formed. The committee's main responsibility is to establish the scope and content of the program. Since the stakeholders are predominantly industry practitioners, they provide assurance that the program will be job-related, realistic and technically current. The selection of the committee members is made with a viewpoint toward ensuring national representation and diverse employer type and size. Each committee is normally composed of about ten persons who are well versed in the identified technical area and can be accepted by others as Subject Matter Experts.

The Practice Analysis

The first objective for the committee is to conduct a practice analysis and develop a profile delineating the typical responsibilities and experience of the persons to be certified by the program. The profile is usually drafted by one of the committee members and then reviewed and refined by the entire committee. It depicts the career path of the technician from entry level (level one) to the senior level (level four) in the particular technical specialty area. It describes minimum work experience, on-the-job responsibilities, typical activities, and typical job titles of the technicians for each of the four certification levels in the specialty area (Table 1).

The profile is an important document which serves to identify the individual for whom the certification program is being developed. It also serves as the foundation and guide for the development of the work elements, and becomes a part of the program documentation that is made available to the public.

The Work Element Matrix

Development of the profile is followed by assembling a work element matrix for the program. The matrix serves as an aid in the development and classification of the work elements. It has four levels of work elements as its horizontal axis and the categories for the elements as its vertical axis. The categories represent the domains or general subject areas within which related work elements are grouped at the particular levels.

The use of categories in the development phase insures that all the appropriate subject matter is covered and eases the necessary comparing and contrasting of similar

Table 1 – *Potential Technician Profile For Certification In Hot-Mix Asphalt Construction*

	Level I	Level II	Level III	Level IV
Education	There is no formal education requirement. (Note 1)			
Minimum Work Experience	Limited experience in bituminous materials QA/QC or related activities.	Minimum of two years, of which at least one year must involve bituminous materials QA/QC activities. The balance (12 months) may be related activities in any capacity or other related specialties such as constuction inspection, surveying, etc.	Level II work experience plus three additional years. At least three of these years must involve bituminous materials QA/QC as the primary activity. The balance (2 years) may be construction related activities in any capacity.	Level III work experience plus five additional years of bituminous materials QA/QC experience involving a broad range of complexity and diversity.
Level of Responsi- bility	Assignments requiring direct supervision.	Routine QA/QC assignments performed under general supervision.	Bituminous materials QA/QC work performed with minimal supervision. May provide daily supervision for one or more persons who are Level I or II.	Independent bituminous materials QA/QC work including delegated responsibilities and duties for which engineering precedent exists. Assign tasks/ supervise personnel.
Typical Activities	Perform simple tasks, measure- and computations. Document findings.	Perform common acceptance tests. Monitor bituminous materials construction procedures. Prepare project test reports.	Conduct common and specialized tests. Monitor common and unique construction procedures. Verify locations and quantities. Maintain records. Offer recommendations.	Project management, oversee specialized tests and complex construction procedures. Interact with project engineer/manager. Recommend corrective actions.
Typical Job Titles	• Trainee	• Laboratory Technician • Inspector • QC/QA Technician	• Lead/Senior Technician/ Inspector • Senior QC/QA Technician • Lab/Field Services Representative	• Chief Technician/ Inspector, • Chief Lab/Field Services Technician • QA/QC Manager

Note 1: Certification at Level II and higher assumes prior educational experiences (college, self-study, correspondence courses, workshops, etc.) that develop knowledge equivalent to that achieved from courses in a civil engineering technology or closely related associate degree program.

work elements which occur at the different levels of the matrix. After the program becomes operational, the categories are used in identifying those work elements which might serve as "crossovers" for similar work elements in other NICET certification programs.

The categories are loosely defined by simple titles. The following are typical category areas in which testing might be appropriate in a typical materials engineering technician certification program:
- Communications
- Mathematics
- Science
- Regulations, Standards, and Practices
- Terminology
- Equipment Type and Use
- Processes
- Sampling, Testing and Measurement
- Data Analysis and Reporting
- Work Site Safety
- Work Management

The above listing is a starting point for the committee. The final list for a specific certification specialty area may contain more or less categories.

Work Element Development

Using the assembled matrix, an initial, "rough" set of work elements is then generated for the various categories and levels.

Each work element consists of one or more logically connected tasks or related knowledge units. The tasks which make up a work element must have some common connection, such as use of a piece of equipment, or a sequential execution of a process, etc. Accordingly, a work element has a specific, observable objective such as a product, a specified accomplishment, a synthesis of information, etc.

An important criterion for inclusion of a work element in the program is its importance relative to other identified work elements. As all work elements carry the same weight in terms of satisfying certification requirements, they must also have roughly the same importance as a measure of technician competence. Questions which should be considered during the selection of work elements are:
- Are there any work elements which are not important andcan be removed without decreasing the value of the certification?
- Are there any important tasks or knowledge units which are not already included and which would increase the value of the certification?
- Can non-trivial questions be written for each work element which can be answered by competent technicians?
- Is a task or knowledge unit currently being covered by two or more work elements when it could be better handled by a single work element?
- Should a work element be split to better cover the tasks and/or knowledge involved?

Work elements are developed for each of the four certification levels to reflect a technician's progression from very basic skills in some job areas to a high level of competence in most areas of the specialty and an ability to deal with complex problems and situations. Distinction between job tasks which require the technician to work with a more knowledgeable person, work without supervision, or supervise others is also recognized by locating work elements at the various levels within the program.

Table 2 - Possible Examination Requirements

Work Element Category	Number Of Program Work Elements	
	Required For Certification	Available For Testing
I GENERAL	5	9
I SPECIAL	4	13
Total:	9	22
I GENERAL	7	9
I SPECIAL	6	13
II GENERAL	6	10
II SPECIAL	10	30
Total:	29	62
I GENERAL	7	9
I SPECIAL	8	13
II GENERAL	7	10
II SPECIAL	18	30
III GENERAL	11	14
III SPECIAL	4	6
Total:	55	82
I GENERAL	7	9
I SPECIAL	10	13
II GENERAL	7	10
II SPECIAL	25	30
III GENERAL	11	14
III SPECIAL	4	6
IV GENERAL	6	8
IV SPECIAL	1	2
TOTAL:	71	92

Work Element Classification

At each level, work elements are further categorized as "Generals" or "Specials". "Generals" may include one or more "Core" work elements as needed. The "Generals" are work elements which are typically mastered by all technicians working in the specialty area regardless of the place of employment or the type of employer. "Core" work elements are those which are considered to be so essential that they are required of every certificant. "Specials" are work elements that will be mastered by some, but not all, technicians working in the specialty area. Regional conditions and practices and the scope of the employers" business, policies and practices will dictate exposure, or lack of exposure, to the tasks associated with special work elements.

A complete set of work elements properly identifies all important job tasks and knowledge normally encountered by technicians working in the specialty area. Furthermore, the total number of work elements in each classification must be sufficient for testing purposes. The exam requirement for certification at a particular level consists of passing a specified number of work elements from each of the relevant classifications. A typical program would have a total of 80 to 130 work elements, with 70-75% being required of individuals who eventually certify at level IV (Table 2).

The goal is that 85 to 90% of the "General" work elements at a given level will be selected, tested and passed by an individual who certifies at that level. The exception to this is at Level I where a lower percentage is usually required. Additionally, for each level there should be, at a minimum, at least two extra "General" work elements available beyond the total number of "General" elements required. For example, if the total number of "General" work elements available is 9, then the examination requirement for certification would be 7 of the 9 work elements. On the other hand, the goal for the "Special" work elements is that about 70 to 80% of the available work elements will be tested and passed by the time the highest level of certification is reached. The percentage required increases as the technician progresses from the initial level of certification to the upper (senior) level of certification, thus reflecting broader exposure to these specialized tasks. For example, if the total number of Level II "Special" work elements available for testing is 30, then 10 may be required for certification at Level II, 18 for Level III and 25 for Level IV (Table 2).

The NICET maximum testing time for an exam sitting is 7 hours. Therefore, NICET currently allows a maximum of 34 work elements to be tested by an individual on a given day. The goal is to design the certification program such that the number of work elements required for certification at Level II, as well as the number required for upgrading from Level II to Level III and from Level III to Level IV is between 25 to 28 so that the exam requirement for each level can be completed in one day of testing. This also offers the examinee the flexibility to select and test a number of work elements beyond the minimum requirement which is specified for each work element classification.

Work Element Descriptions

Work element descriptions serve as a guide to the individuals who are selecting work elements on which to be tested and to those who are gathering reference materials for review and use during the open-book written examination. Therefore, the descriptions have to be as clear and unambiguous as is possible. However, they must also convey enough specific information to both the person who has already acquired and the person who has not yet acquired the requisite skills and experience to understand what competencies will be evaluated. Technical standards are referenced for each work element as appropriate. References must be nationally recognized and readily available.

Committee Review

A full review of the work element titles, descriptions and technical references is then conducted by the committee members. These members review the list to judge whether the set is complete and if each category and level are adequately covered. They also check to see that titles, descriptions and references are an accurate representation of the scope, content and skill level of work done by the candidates for certification. A refined listing and matrix is then prepared for a review by a sample of the industry as a whole.

Program Validation

The review by industry stakeholders is extremely important because it is really a validation of the contents and purpose of the certification program. Among those asked to critically review the program will be engineering technicians, supervisors, trainers, engineers, and, if appropriate, regulators and educators. They are asked to review, comment and rate the work elements, both individually and as a collection. Based on these evaluations, the committee determines appropriate revisions and generates a "working" version of the program's work elements which are then used for question development and a final validation through actual field testing.

Work Element Topics

Each work element is divided into a set of topics which delineate subject areas for which exam questions are written. A topic is a discrete action, task or knowledge unit, mastery of which is important in determining work element competency. A complete set of topics assures that the questions in each work element question bank will cover the range of important/critical activities described in the work element, rather than concentrating on one or two sub-areas. The topics are written with the question writing process in mind. This involves close scrutiny of the practical meaning of the work element itself, and often leads to improvements and enhancements of the actual work element descriptions.

The number of topics in a work element equals the number of questions to be assigned to one bank for that element. Each work element will typically have five topics. It may, on occasion, be appropriate for an element to contain more than five topics. If more than ten topics surface, consideration is given to splitting the work element into two elements.

Three question banks are usually developed for each work element. Thus each topic must "generate" a minimum of three "equal" but "different" questions. These banks allow different sets of questions to be presented to individuals who fail and retest the work elements. This arrangement also allows quick intervention if a question bank is ever compromised.

Question Development

An important component of the NICET model is the multiple-choice written examination. The written exam is designed, developed, and administered in accordance with industry standards that are driven by psychometric and legal issues. It is an objective, economical and fairly reliable measuring device. The logistics required for its proper administration are relatively easy and offer much flexibility. Scoring and documentation are straightforward, defensible and can be easily filed for future reference.

Questions are written in the "multiple-choice" style, and range from simple memory types to ones requiring more complex multi-step processes to others involving very complex calculation, analysis and "best action" decision making. All questions are tailored to the type and level of the individual work elements. Submitted questions are edited and then evaluated by subject matter experts other than those who initially wrote the questions. Based upon this review, questions are "finalized" and then field tested by a group of technicians who are representative of the potential pool of candidates for the new certification program. After field testing, a comprehensive review of question performance is conducted. Necessary adjustments are made by the committee after which the program becomes operational.

The Program Detail Manual

The final step in the program development process is the preparation and production of a program detail manual. The manual contains information needed to apply for the certification examinations, including, general information and procedures, a technician profile, an examination requirements chart, certification requirements, a listing of work elements and descriptions, and selected general references. This program detail manual is to be used by each and every technician seeking certification in the specialty area.

Candidate Evaluation and Certification

Once a candidate satisfies a specific examination requirement for one or more of the four certification levels in a particular specialty area; work element verification, work history and other provided documentation are evaluated for compliance with the requirements of the program.

NICET's method for the evaluation of on-the-job performance is through a process by which the supervisor confirms that the candidate has demonstrated competence in the work elements being tested. The verifier must have the technical expertise in the specialty area and must have first-hand knowledge of the candidate's specific job skills. Each verifier provides NICET with details of her/his employer, job title, qualification and the nature of her/his relationship with the candidate. The verifier also signs a statement of understanding to the fact that by verifying the candidate's work elements, he/she is certifying that he/she has observed the applicant repeatedly and correctly perform the tasks or utilize the knowledge required in the specific work element under a variety of conditions. This NICET on-the-job performance evaluation process is simple and economical, but occasionally it is jeopardized by a smaller number of verifiers who are not as honest as others in the profession expect them to be. NICET does scrutinize verifiers and when improper verifications can be established, NICET can and does permanently deny the certification sought or revoke the certification(s) held by both the applicant and the verifier (if certified).

Other forms of performance evaluation exist, such as oral and practical examinations,

but each has its inherent disadvantages. Performance exam development and administration can be expensive and complicated, and the scoring process will be more subjective than that for the multiple-choice written exams. Important questions which should be asked when administering such exams are:
- Is the behavior being measured something that could not be evaluated by the use of a multiple-choice or objectively scored examination?
- Are the evaluators thoroughly trained prior to the examination administration?
- Are there detailed criteria for evaluating and scoring?
- Does each evaluator make an independent rating?
- Are at least two independent evaluations made for each candidate?
- Is the evaluation free of potentially biasing information about the candidate which is not related to examination performance?
- Has the examination session been documented (proctored, audio or video taped)?

Another consideration is that all forms of examinations should be standardized so that all candidates have the same opportunity to demonstrate competence. Dissimilar forms of an examination and/or dissimilar testing conditions can quickly result in legal action by the candidate against the certifying body for improper discrimination.

The certification criteria used by the NICET model provides an efficient but simple process of checks and balances. Each time a candidate meets an examination requirement, his or her file, which includes all application forms and applicable documentation, is completely reviewed for consistency and accuracy of job details, relevancy of work history, progression of duties and responsibilities, etc. The NICET evaluation process allows the discovery of inconsistencies between examination performance, work history, and supervisor verifications, and sets in motion a process which gathers additional information which will influence current and future decisions to award or not to award certification.

Certification, therefore, is not an automatic outcome of meeting an examination requirement. In a typical pool of NICET technicians testing in a cycle, about 60% meet an examination requirement, out of which only about 50% meet all the requirements and are certified. The other 50% are questioned in regard to their work history, work element verification, etc.; and are not certified until the candidate does what is necessary to overcome the deficiencies. Many candidates require months, and sometimes years, to rectify the deficiencies.

Recertification

Certification philosophy is shifting from a credential issued for life to a credential that is issued for a finite time period; and is reissued only if the certificant engages in professional development. All NICET certificants have been assigned an expiration date for their certifications; and during 1999, the vast majority of them will be applying for recertification.

The NICET recertification period is three years; and recertification is possible only if the certificant has accumulated, during the prior three years, 90 Continued Professional Development (CPD) points for each certification held. These 90 points are accumulated through activities in the categories of Active Practitioner, Additional Education, Advance Profession, Certification Activity, and Special Exam. An important condition is that at least two of the five categories must be used to accumulate the 90 CPD points.

NICET's experience to date with recertification is limited. However, our expectation is that a number of certificants will be giving up their certification altogether or reducing the total number of certifications held because of inactivity in a specific certification area. This is a fitting outcome since most end users of any credential expect the holder to be current and actively involved in the concerned specialty area.

Summary

The NICET model for creating job-task competency based certification program begins with a practice analysis conducted by a technical advisory committee to identify the important job tasks performed by engineering technicians employed in the identified technical specialty area together with the time period during which the technicians should become proficient in these tasks. The outcome of the practice analysis and subsequent industry validation is a collection of work elements and specific requirements for certification. Collectively, the work elements cover the knowledge and skill required of engineering technicians working in the specialty area. The program design emphasizes the career path and progression of the technicians from entry levels all the way to senior and supervisory levels. Having the work element as a "stand alone" module with its own examination questions allows changes in technology, work practices and standards to be easily accommodated.

The model offers benefits to all the stakeholders. Technicians benefit from 1) a reliable evaluation of their technical strengths and weaknesses; 2) a third-party acknowledgment of their acquired knowledge and abilities; and 3) nationwide recognition as a qualified professional. Owners, managers and supervisors benefit from 1) a relatively quick, efficient and economic method to identify the specific training needs of their technicians; 2) evidence that they employ qualified personnel; 3) a nationally applicable certificate; and 4) a marketable advantage over their competitors.

References

[1] "Certification: A NOCA Handbook," A. H. Browning, A. C. Bugbee, and M. A. Mullins (Eds.), The National Organization for Competency Assurance, Washington, DC, 1996.

[2] "Development, Administration, Scoring, and Reporting of Credentialing Examinations: Recommendations for Board Members," Council on Licensure, Enforcement, and Regulation, The Council of State Governments, Lexington, KY, 1993.

[3] "Program Detail Manual for Certification in Construction Materials Testing: Asphalt, Concrete and Soils," The National Institute for Certification in Engineering Technologies, Alexandria, VA, (www.nicet.org), 1997.

Deborah G. Hutti [1] and Larry Hymes[2]

Lake Land College / Illinois Department of Transportation: Quality Control / Quality Assurance Training Program – Development and Implementation

Reference: Hutti, D. G. and Hymes, L., **"Lake Land College / Illinois Department of Transportation: Quality Control / Quality Assurance Training Program – Development and Implementation,"** *Hot Mix Asphalt Construction: Certification and Accreditation Programs, ASTM STP 1378*, S. Shuler, and J. S. Moulthrop, Eds., American Society for Testing and Materials, West Conshohocken, PA, 1999.

Abstract: Since 1992, Lake Land College (LLC), in cooperation with the Illinois Department of Transportation (IDOT), has employed a nontraditional approach to an ongoing need within the State of Illinois for Quality Control / Quality Assurance training. LLC's Quality Control / Quality Assurance initiative utilizes the educational and training resources of a rural community college to fulfill a need within the Illinois quality management program.

This paper will discuss the development and implementation of LLC's Quality Control / Quality Assurance training program from the initial inquiries about a potential program at the College in 1992 to the success it has maintained today. The information will be detailed using a chronological approach and will focus on academic, industrial, and governmental collaborative success. The outlining of this triangulation will provide a roadmap for the development of similar programs in the area of Hot Mix Asphalt Construction: Certification and Accreditation Programs.

Keywords: quality control / quality assurance, Illinois Department of Transportation, Lake Land College, quality management program, community college, hot mix asphalt, Illinois Department of Transportation (IDOT)

---

[1] Coordinator, Central Illinois Distance Education Network, Lake Land College, NW30, 5001 Lake Land Blvd., Mattoon, IL 61938.
[2] Director, Quality Control/Quality Assurance, Lake Land College, 5001 Lake Land Blvd., Mattoon, IL 61938.

Overview

Since 1992, Lake Land College (LLC) has worked in conjunction with the Illinois Department of Transportation (IDOT) to develop and implement a training program that meets the quality control/quality assurance guidelines in the area of Hot Mix Asphalt that have been set by the State of Illinois. In the early 90s, IDOT's Bureau of Materials and Physical Research established a state-wide quality management program for contractors involved in road paving. At that time, due to LLC's on-going collaboration with business and industry within their district, LLC and Howell Asphalt Co. (a local HMA industry), approached IDOT and inquired about the possibility of offering the mandated quality management training through the community college system, in particular through Lake Land College. In prior years, Lake Land College's Division of Civil Engineering Technology had developed a successful track record with Howell and with other similar industries throughout the state in preparing students with skills necessary for employment as civil engineering technicians with consulting firms, testing laboratories, utilities and local, state, and federal agencies. Simultaneous with Howell's request, IDOT realized that the number of individuals who were in need of the mandated training was much larger than what IDOT could provide, and they began to pursue the idea of contracting with a community college as a provider. Eventually, Howell Asphalt partnered with LLC, providing access to equipment, technicians, and expertise which enabled the College to successfully bid on the IDOT's quality management training contract.

The initiative utilizes a community college with expertise in the area of teaching and learning to fulfill a need within the state's quality management program. It uses a public educational institution to provide government mandated training. All individuals and industries involved with Hot Mix Asphalt and Portland Cement Concrete contracts within the State of Illinois are responsible for sampling, testing, and documenting for specification compliance (Quality Control). IDOT is responsible for random monitoring testing (Quality Assurance). Lake Land College is the sole provider of Quality Control / Quality Assurance training necessary for individuals and industry active in specification compliance and IDOT employees active in testing. LLC offers ten courses at two locations throughout the state – Howell Asphalt Co. in Mattoon, and Gallager Asphalt Co. in Thorton. In order to provide training at the site in Thorton, LLC has communicated and collaborated with other community colleges in the Chicagoland area, partnering with Prairie State College as a secondary site that allows for out of district training. The QC/QA program provides participants with the ability to:

- Administer and complete the testing required for an aggregate producer participating in the Aggregate Gradation Control System,
- Complete the testing associated with contracts let under the QC/QA Program,
- Manage a Quality Control Program for contracts let under the QC/QA Program,
- Complete a design mix for Hot Mix Asphalt,
- Understand the use of the nuclear density gauge, a requirement for individuals running nuclear density on asphalt QC/QA projects,
- Meet safety training requirements for IDOT for individuals operating Nuclear Density and Nuclear Asphalt Gauges,

- Understand the use of the Superpave Gyratory Compactor and Ignition Oven for QC/QA during the production of Hot Mix Asphalt,
- Complete the mix testing for Portland Cement Concrete on QC/QA projects and receive ACI Level I certification, and
- Understand the proportioning of Portland Cement Concrete for QC/QA projects.

The QC/QA Partners

Lake Land College

Lake Land College (LLC) is a comprehensive community college with a proud tradition of academic excellence through quality educational programs. It is located on a 307 acre campus in rural east central Illinois, serving the second largest geographical community college district in the state. LLC spans over a vast 4,000 square mile area which covers 36 school districts within all or part of 15 surrounding counties. The total population of the district is approximately 180,655 with the majority of its residents living in small rural towns. The student population is slightly over 5,500 with more than 67% of all in-district high school graduates who enroll in college attend LLC.

LLC is dedicated to the principle that learning is a lifelong process and that education and training are essential to a free and open society. We encourage vision, leadership, and action. Meeting the ever changing training and employment needs of individuals, businesses and organizations is Lake Land College's ongoing objective. Providing quality employees who emphasize the individual attention, advanced technology, services, and accommodation to student needs is our responsibility. Lake Land College is an open door post-secondary institution where all students qualified to complete any programs of their choice are admitted and advised. Cultural enrichment, social activities, and learning opportunities and services are provided to the community serviced by the College.

Howell Asphalt Company

Howell Asphalt Company in Mattoon, IL and was founded in 1951 by Virgil R. Howell. It is a regional contractor specializing in asphalt production and paving in east Central Illinois. It is a third generation family-owned company, currently led by Mr. Charles Adams and Mr. Ken Ozier. The Howell Asphalt Company has received the National Asphalt Pavement Association – Quality in Construction Award, the Federal Highway Administration – National Quality Initiative Achievement Award and several IDOT-Bituminous Concrete Pavement Awards of Excellence.

Gallagher Asphalt Company

Gallagher Asphalt Company is also a third generation family owned company and operates in the Chicago area. The states of Illinois and Indiana have recognized Gallagher's dependability track record by awarding a 99%+ Rating for both Quality Assurance and Early Project Completion. Because of their high quality work, Gallagher Asphalt Company has received dozens of awards including Consideration of the Public,

Quality, Assurance, Performance, and Value Engineering. The leadership of Gallagher has been active in industrial associations, in particular the Illinois Asphalt Pavement Association and the Illinois Roadbuilders Association.

Other Participants

The Illinois Asphalt Paving Association serves in an advisory capacity to LLC's QC/QA program. They assist in the development and implementation of courses. The Asphalt Institute assisted specifically with the development of the Level III Mix Design course and furnished the instructor for that course for several years.

Chronological History of the QC/QA Project

1990-1992

Prior to 1990, the IDOT had employed the traditional method of controlling quality, with IDOT employees completing all inspection, testing, and decision making in regards to managing quality. In 1990, IDOT joined the nationwide movement towards QC/QA, let by Mr. James Gehler, the Chief of the Bureau of Materials and Physical Research. The decision was made by IDOT to include not only hot mix asphalt, but concrete and aggregates as well in their QC/QA program. IDOT's early activities included assembling the QC/QA training manuals, developing curriculum, staffing the courses, and providing professional development for those teaching each individual course. In addition, IDOT developed all the specifications for administration of the QC/QA contracts. In the winter of 1991, they offered the first courses for their contractors at the Bureau of Materials and Physical Research offices in Springfield, Illinois, taught entirely by state personnel and offered free of charge.

1992

In 1992, LLC Civil Engineering Technology (CET) program had been in continuous operation for over twenty years. At the time, it was one of the few programs of its type within the state of Illinois and had been acknowledged by IDOT as a quality civil engineering technology program. Through LLC's Associate in Applied Science degree, the CET program prepared students with skills necessary for employment as civil engineering technicians for consulting engineering firms, testing laboratories, utilities, and local, state, and federal government agencies. Emphasis had been and still is placed on surveying, material testing, drafting, and construction inspection associated with civil engineering. Because the CET program had graduated approximately twenty students per year for over twenty years, by 1992 a significant number of LLC graduates were currently employed in the highway industry, both with contractors and IDOT. In fact, Mr. Jack Davis, an LLC graduate, was one of the instructors who was employed by IDOT to develop and teach one of the advanced asphalt courses in Springfield. In addition, several other graduates were currently employed by Howell Asphalt Company.

In December of 1991, Mr. Adams of Howell Asphalt approached LLC about the possibility of developing an alternate method of QC/QA training, one that would be more localized and accessible for their employees. The initial idea was to use the teaching and learning expertise of LLC to provide QC/QA training for Howell Asphalt employees and other individuals within LLC's district.

In January of 1992, LLC, Howell Asphalt, and IDOT entered into a series of meetings which produced multiple, pivotal observations on the idea of alternate sources of QC/QA training outside of the state agency. In attendance at these meetings were administrative employees from IDOT, the CEO of Howell Asphalt, LLC's Associate Vice President of Workforce Development, and Mr. Larry Hymes, who was currently the Civil Engineering Technology Program Director. The observations included: 1) IDOT did not want to fragment the training. Regardless of the source, they wanted to have one singular source of training for the entire state in order to insure uniformity. 2) They had to consider the cost effectiveness of maintaining the current method of training, expanding their site and continuing to provide QC/QA services without change as opposed to outsourcing it. 3) They needed to consider all alternate educational sites, including other community colleges and four year institutions within the state who would be capable of providing training and who may be interested in the program.

Throughout these early meetings, LLC was able to garner a clear picture of the type of program that would be of interest to IDOT, and within several weeks, the College submitted a proposal for IDOT to consider. Partnering with Howell Asphalt, LLC agreed to provide training for the entire state through a tuition based program. All program participants would pay tuition, including employees of the state. During the program's first year, Howell Asphalt agreed to loan various pieces of equipment to LLC and also agreed to furnish technicians to help the College with instruction during peak enrollment periods. These periods generally occurred during the winter months, which have been traditionally slow periods for the industrial partner's technicians. The partnership with Howell Asphalt allowed LLC to enter into a new training program, one that is a labor intensive and equipment intensive operation in a fiscally responsible manner.

In the summer of 1992, LLC was awarded a three year contract to provide QC/QA training. At the same time, IDOT awarded six construction contracts, of which Howell Asphalt received two. In order to develop an operational program by Fall, LLC focused on several main areas: the location of the lab, the securing of the equipment, the registration procedure, and the instruction process.

Location of Lab – Initially, LLC looked at various off-campus sites as locations that would fulfill an immediate and pressing need. And although off-campus sites satisfied the need in the short term, LLC recognized that in the long term, an on-campus site would provide easy access to the administrative services of the College needed by the program. Thus LLC began to renovate an existing building which has become the QC/QA permanent location.

Securing of Equipment – Those developing the equipment list were able to secure information from IDOT concerning equipment used within their training program which provided a basis for identifying mandatory equipment needs for Fall. At the same time, LLC began coordinating the procuring of this equipment through available funds and through the partnership with Howell Asphalt and IDOT. As it turned out, most of the equipment used during the first semester was loaned to LLC from its partners.

Policies and Procedures – As the partnership with IDOT was unique to the LLC community, it was necessary for those involved in the program to develop their own policies and procedures concerning admission, enrollment, student services, bookkeeping, registration, and data collection. These policies and procedures still had to blend with those at LLC, but they had to also satisfy the recommendations from IDOT. In order to coordinate all of the activities associated with policies and procedures, however, LLC QC/QA program engaged the services of a part-time secretary. All day-to-day clerical needs were handled through this position.

Instruction Process - During the Fall of 1992, IDOT agreed to provide professional development for those who would be teaching QC/QA courses at LLC. IDOT experts collaborated with the one instructor hired at LLC to provided the QC/QA training and technicians from Howell Asphalt who would be employed intermittently in the instruction process in mentoring these individuals through the curriculum. IDOT agreed to continue the mentoring process throughout the beginning of the first courses in order to make sure that the curriculum taught blended well with their original plan. Without the IDOT mentoring process, it is doubtful that the instructor and technicians would have been prepared to teach courses as soon as they actually did.

In November of 1992, LLC offered its first series of QC/QA courses, and found that the enrollment was approximately twice what had been projected. Initial projections were 288 with enrollment reaching 610. In addition, the students who evaluated the training indicated a high level of satisfaction. They were appreciative of the fact that the instructors within the program had a tremendous amount of experience within the industry, and that the program seemed to run smoothly due to its affiliation with a institution of higher education.

First Year Problems – LLC's initial QC/QA program, however, was not without problems. First, the length of time between the contract agreement with IDOT and the offering of classes was too short to allow for the complete preparation of facilities. Though appreciative of the extensive work that had been completed from August to November, the QC/QA training site was far from finished. Though the building will always require small ongoing modifications, it was completed by Spring of 1993, a span of approximately eight months.

A second problem had emerged by the end of the first training class. The enrollment was nearly double what had been expected which demonstrated a clear need for a second full-time instructor as well as a second designated classroom. In addition, the part-time secretary was unable to keep up with the day to day clerical needs.

1993

In August of 1993, LLC hired a second full-time instructor, Mr. Galen Altman. He had over ten years of experience in an IDOT district, conducting a portion of their asphalt testing in the materials lab. At LLC, he would staff classes that were added due to the increased enrollment. At the same time, LLC began renovating the QC/QA building to add a second classroom. The two QC/QA designated classrooms would share the existing equipment lab, which had been redesigned and resupplied with items identified and specified throughout the first classes. A majority of the equipment was still on loan from Howell Asphalt and IDOT; however, little by little, LLC was purchasing pieces of equipment and returning items to their partners.

In order to eliminate the clerical difficulties that had emerged due to lack of sufficient human resources, a student programmer was hired to 1) develop a computer program to assist in managing the enrollment, and 2) to complete routine data input tasks that could not be finished by the part-time secretary due to lack of time.

By the beginning of the Fall 1993 semester, renovations on the second classroom had been completed and due to the immediate success of the program, LLC was able to purchase the majority of the equipment needed and to return all of the equipment which had been borrowed from Howell Asphalt and IDOT.

In order to satisfy the rising enrollment, LLC began teaching some of the lecture-only classes at the College in DuPage, located slightly west of the Chicago Metropolitan area.

Second Year Problems – It was obvious by the end of 1993 that although the two classrooms provided adequate space for the current enrollment, the sharing of the equipment lab caused scheduling difficulties. At times, the instructors had to schedule classes on Saturdays in order to provide their students with adequate lab time to complete the QC/QA training.

By the end of 1993, it was also clear that a need existed for courses to be offered by LLC in other parts of the state. Enrollment had increased by 14% over the past year; however, LLC was still not serving a portion of individuals, 1) who had requested QC/QA training but were located too far away from the College to be able to complete the training, or 2) who were unable to register for the most convenient course time and section due to the overwhelming enrollment within the program.

And, as experienced during the previous year, it was found that the part-time secretary and the student programmer could not keep up with the amount of QC/QA clerical needs.

1994

During the summer of 1994, the QC/QA process at IDOT had evolved to the point where practically all of the asphalt work in the state of Illinois, with the exception of District 1, the Chicago area, was QC/QA; thus, LLC was well aware of the potential for an increase in enrollment throughout the next year. The two instructors, two classrooms, and one lab seemed adequate; however, in order to eliminate the continuing difficulties with the policies and procedures surrounding enrollment, admission, registration, fee collection, bookkeeping, and data collection, Ms. Marlene Browning was hired as a full-time office manager. Because of her full-time status, she was able to eliminate the difficulties faced in this area since the inception of the program and provide significant information concerning the financial status of the entire program.

During the remainder of 1994, there were no significant new developments within the program. The partnership among IDOT, LLC, and Howell Asphalt had been running smoothly, and the program at LLC had not encountered any of the difficulties experienced in previous years. As LLC entered into the last year of its current three year, renewable contract with IDOT, the following courses had been developed and offered (see Table 1 below):

Table 1: *QC/QA Courses Offered*

Course Title:	Year Offered
3-Day Aggregate for Mixtures	1993
5-Day Aggregate Technician	1993
Level I Hot Mix Asphalt (5-Day)	1993
Level II Hot Mix Asphalt (5-Day)	1993
Level II Portland Cement Concrete (2-Day)	1993
Level III Hot Mix Asphalt (5-Day)	1994
Nuclear Density (1/2-Day)	1994
Nuclear Safety Course (1-Day)	1994
Superpave Field Control (2-Day)	1997
Level I Portland Cement Concrete (3-Day)	1997

1995 - 1996

In the summer of 1995, LLC was awarded a one-year contract extension with the understanding that at the end of this period of time, it would be necessary to present IDOT with a multi-year proposal. IDOT wanted a long-term commitment from their training contractor, especially as the Chicago area began joining the movement towards QC/QA and as the Superpave technology began to be implemented. In the summer of 1996, IDOT issued a Request for Proposals for the QC/QA training that included two new requirements. The training contractor would be responsible for maintaining the QC/QA manuals, establishing a Chicago-based training facility, and conducting classes in the Superpave system, which would involve buying Superpave Gyratory Compactors, Binder Ignition ovens, etc. The RFP stipulated that the contract would be valid for two years with the potential for an additional two years after a review by both parties. LLC submitted a proposal and IDOT awarded the contract to the College.

As soon as the contract was awarded the full-time office manager began preparing to produce the training manuals. This particular activity required additional coordination among the manual contributors and additional hardware and software purchases.

In order to fulfill the requirement to provide Chicago based training, LLC generated a list of all the potential contractor partners in that area. Through the expertise of the QC/QA faculty and through conversations with others who had experience in the Chicago area, LLC contacted the Gallagher Asphalt Company in Thorton, Illinois as a possible partner. In a meeting between Mr. Gallagher, CEO and President of the company, and LLC QC/QA program members, it was decided that Gallagher Asphalt would build a lab and lease it back to LLC. In turn, LLC would train and hire a minimum of two Gallagher technicians to participate in the instruction at this site. During the winter, it was necessary for one LLC instructor to teach courses in Chicago, aided by the Gallagher technicians. Again, enrollment increased (see Table 2 below)

Fifth Year Problems – As the number of classes offered increased once again, it became increasingly difficult for a LLC instructor from Mattoon, IL to teach classes in the Chicago area. (During the winter alone, LLC QC/QA counted more than 70 overnight stays in the Chicago area). It was clear to all that a third instructor had to be hired to staff the Chicago location.

Table 2: *QC/QA Enrollment Summary*

Course	FY93	FY94	FY95	FY96	FY97	FY98	Total
Aggregates for Mixtures	177	197	237	202	216	224	1253
Level 1 HMA	173	159	185	173	163	179	1032
Level 2 HMA	136	124	127	118	107	110	722
Level 3 HMA			76	70	50	40	55	291
Superpave Field Control					95	88	183
Total	486	556	619	543	621	656	3481
% Change		14	11	(12)	14	5	

1997 - 1998

During the summer of 1997, the search for a person who had experience with the QC/QA program, specifically in the materials area, began. After a lengthy search period, LLC hired Ms. Lori Walk, as the Gallagher site instructor. At the same time, the Gallagher Asphalt Company decided that due to the success of the project, they would spin off the training center as a second company. The new company, JFG Technical Center, offered a variety of services to its customers beyond the QC/QA training, including: materials testing, materials consulting, and QC or QA on a for-hire basis. This company began operating as a separate entity in January of 1998.

The current year is bringing additional changes as IDOT is in the process of expanding the QC/QA concept to local agencies. At this time, QC/QA is used for state contracts only. This expansion will require new course work and new training. In addition, there is a movement to expansion into the aggregate industry. At this time, only the aggregates used in asphalt and concrete are included in the QC/QA program. IDOT hopes to expand and include all aggregates, as well as natural sands and gravels.

As IDOT becomes more and more comfortable with LLC's QC/QA program, it becomes more and more apparent that the College's program will only continue to grow.

Serji N. Amirkhanian[1]

South Carolina's Experience with Certification and Accreditation

Reference: Amirkhanian, S. N., **"South Carolina's Experience with Certification and Accreditation,"** *Hot Mix Asphalt Construction: Certification and Accreditation Programs, ASTM STP 1378*, S. Shuler, and J. S. Moulthrop, Eds., American Society for Testing and Materials, West Conshohocken, PA, 1999.

Abstract: Approximately five years ago, the SC DOT started the certification process for their asphalt program. The DOT decided to conduct all of its courses and the certification program at Clemson University. Initially, there was only one certification course (i.e., Marshall Method of Mix Design). Presently, there are five different courses, including: HMA QC Technician (Level 1); Marshall Method of Mix Design (Level 2); Superpave Mix Design (Level 2S); QA/QC Manager (Level 3); and Roadway Inspector. Most of these courses are at least four days long. In 1999, fourteen of these courses will be offered. Most of these courses are team taught. Experts from industry, DOT, and academia teach a portion of each course. Based on the comments from the participants, some from other states, the courses have been, in general, a welcome addition to the industry.

Keywords: asphalt, mix design, certification and accreditation, HMA technicians

Introduction

Approximately five years ago, the South Carolina Department of Transportation (SC DOT) initiated the certification process for their asphalt program. There was a need to ensure that all contractors and DOT inspectors are following the proper procedures. The DOT decided to conduct all of its courses and the certification program at Clemson University, Civil Engineering Department. The first program conducted was the Marshall Method of Mix Design. The first course was conducted in October 1993. DOT and contractors' representatives selected the participants for this course. In order to establish the agenda, meetings were held with DOT officials and several contractors. The author was selected as the primary instructor for this course. Based on the author's suggestion, it was decided that this course would be team-taught.

---

[1]Professor, Civil Engineering Department, Clemson University, 110 Lowry Hall, P.O. Box 340911, Clemson, SC 29634-0911.

The first course had sixteen (16) participants. The class was divided into two sections: lecture and laboratory. During the lecture period, the concept of the Marshall method of mix design and the SC DOT's testing procedures were discussed. In the laboratory, the participants were divided into four groups and each designed a compete mix design. The course content and the laboratory activities were evaluated after the course was completed. The recommendations were implemented for the next course. The next course had an enrollment limit of twelve people. Due to the lower number of people, this class was much more productive, therefore, it was decided that all classes would have a maximum enrollment of twelve.

After conducting several courses for the Marshall method of mix design, the SC DOT contacted the author and requested initiation of a second course. This course is entitled "HMA Quality Control Technician Course: Level 1". This course was designed for the plant technicians who are responsible for conducting laboratory testing procedures to ensure the mixture follows the DOT specifications. The first course was conducted approximately two and half years ago. Selected individuals from DOT and the industry were invited to attend the course and evaluate the contents. The materials were modified after the class and offered on a regular basis.

Based on the DOT's request, two years ago, the author initiated three other courses including Superpave mix design, roadway inspectors course, and QA/QC manager course. Presently, there are five different courses including: HMA Technician course (Level 1); Marshall Method of Mix Design (Level 2); Superpave Mix Design (Level 2S); QA/QC Manager course (Level 3); and Roadway Inspectors course. Most of these courses are at least four days long. Table 1 shows the course titles, duration for each course, maximum number of participants, laboratory involvement, and the pre-requisites.

The SC DOT procedures indicate that a contractor or a consultant involved with any DOT project must have certified technicians. At this point, no other certifications from other states or agencies are accepted. The main reason for this criterion is that the SC DOT procedures, like many other states, are different than others; therefore, the technicians must be familiar with the state's specifications. The technicians who have been certified at Level 2 must also be AMRL certified. In addition, the field laboratories must be inspected and certified by the DOT officials. The course contents are reviewed and evaluated at least once per year by the Certification Board. Members of this Board consist of people from DOT, industry, and the author.

Table 1 – *Course Information*

Course Title	Level	Maximum Number of Participants	Duration (Days)	Laboratory Involvement	Pre-Requisites
Roadway Inspector	---	20	3	None	None
HMA Quality Technician	1	12	5	Everyday	None
Marshall Method of Mix Design	2	12	4	Everyday	1
Superpave Mix design	2S	10	5	Everyday	1 and 2
QA/QC Manager	3	12	4	None	1 and 2

Course Descriptions

Roadway Inspector Course

This course was designed for people who are responsible for conducting SC DOT testing procedures on the roadway. The course does not have any prerequisites; however, the contractors are informed that they must have the proper personnel to attend the course. The class starts on Sunday with a dinner and introductions that allows the participants to know each other and also lets the instructors know the level of experience of each person. The classes are conducted on an informal basis. The instructors from industry (equipment manufacturers, contractor representatives, etc.), SC DOT, and the academia present all aspects of the roadway work. Table 2 shows a summary of some of the topics covered in this class. The participants are divided into several subgroups and they are responsible for solving some problems related to roadway within a certain amount of time. The solutions to these problems are discussed in the class. This allows the participants to learn from each other and to see other points of views. The participants must pass a written exam at the end of the class. The passing grade, at this point, is set for 75%. There are some discussions to increase this number to 80% in the near future.

Table 2 – *Typical Topics Covered for Roadway Inspector Course*

Topic Title	Day	Approximate Time (Hours)
Responsibilities of Roadway Inspectors	Monday	½
Paving Equipment (Pavers and Compactors)	Monday	1.5
Asphalt Construction	Monday	¾
Materials: Binder and Aggregates	Monday	1
Mix Design Process	Monday	½
Troubleshooting	Monday	1.5
Segregation	Tuesday	1
Pavement Evaluation	Tuesday	1
Compaction	Tuesday	2
Control Strips	Tuesday	½
Effects of Mix Design & Plant on Performance of the Mix in the Field	Tuesday	1.5
Troubleshooting	Tuesday	1.5
SC DOT Specifications	Wednesday	2
Nuclear Density Gauge	Wednesday	2
Exit Exam (Open Book and Notes)	Wednesday	3

Level 1: HMA Quality Control Technician Course

At this level, the participants are expected to have some laboratory experience. In addition, they are expected to be familiar with some of the SC DOT's specifications. Like other courses, the participants arrive Sunday evening for a dinner, introduction to the class, and to get to know each other. This class is divided into lecture and laboratory periods. In the lecture time, the testing procedures, which a typical plant technician would perform, are covered in great detail. Typically, the participants attend the lectures in the morning and go to the lab after lunch. During the laboratory section, the participants are divided into four groups of three. Each group works together the entire week. They perform laboratory procedures and discuss the results with other groups at the end of the week. Table 3 shows some typical topics covered in this class. In addition, Table 4 indicates the laboratory procedures conducted throughout the week.

Table 3 - *Typical Topics Covered for HMA Quality Control Technician Course*

Topic Title	Day	Approximate Time (Hours)
Responsibilities of HMA QC Technician	Monday	½
Aggregates: 　　SC Aggregate Types 　　Physical Properties 　　Sieve Analysis 　　Testing Procedures 　　Specific Gravities 　　Sampling 　　Introduction to Blending	Monday	4.5
Binder: 　　PG Grading 　　Physical Properties 　　Sampling 　　Safety	Monday	1
HMA Mixtures and Introduction to Mix Designs (Marshall and Superpave)	Tuesday	3
SC DOT Specifications	Tuesday	2
Sampling	Wednesday	1
Potential Problems with Mixes	Wednesday	1
Moisture Susceptibility, Rice Gravity, Air Voids, %VMA, %VFA	Wednesday	5
Indirect Tensile Strength, Tensile Strength ratio, Stability, Flow, etc.	Thursday	5
Exit Exam: Written Part (30% of Grade)	Friday	4
Exit Exam: Oral Part (70% of Grade) Both parts are open book and notes.	Friday	1+

Table 4 – *Some of the Laboratory Procedures Covered for Level 1*

AASHTO or SC DOT Designation (ASTM Designation)	Title
T 248 (C 702)	Reducing Field Samples of Aggregate to testing Size
T 27 (C 136)	Sieve Analysis of Fine and Coarse Aggregates
T 11 (C 117)	Materials Finer Than 75-µm (No. 200) Sieve in Mineral Aggregates by Washing
T 176	Plastic Fines in Graded Aggregates and Soils by Use of the Sand Equivalent Test
T 209 (D 2041)	Maximum Specific Gravity of Bituminous Paving Mixtures
T 166	Bulk Specific Gravity of Compacted Bituminous Mixtures Using Saturated Surface-Dry Specimens
T 245 (D 1559)	Resistance to Plastic Flow of Bituminous Mixtures Using Marshall Apparatus
SC-T-75	Extraction Using the Ignition Oven Method
SC-T-70	Laboratory Determination of Moisture Susceptibility Based on Retained Strength of Asphalt Concrete Mixture

On the last day of the class, the participants take a two-level test. They first take a written and then an oral test for the laboratory procedures. These tests are open book and notes. In the laboratory, they must demonstrate that they have mastered the testing procedures based on DOT specifications. The written part of the exam counts for 30% of the grade and the laboratory section for 70%. The passing grade for this course is 75%.

Level 2: Marshall Method of Mix Design

The participants must be Level 1 certified to attend this course. The class starts on Sunday and ends on Thursday. All of the testing procedures shown in Table 4 are covered in this course. In addition, the mix design procedures outlined by the SC DOT are covered in great detail (Table 5). The twelve participants are divided into four groups and each is responsible for a complete mix design including batching the aggregates and performing all the necessary testing procedures. Several other topics such as Superpave design and polymers are also briefly covered in this class. The class is divided into two sections: lecture and laboratory. However, the majority of the time, the participants are working in the laboratory preparing and testing the samples. On the last day of class, a written exam is administered (approximately 4 hours). The passing grade for this course is 80%.

After passing the exam, within the next two months they must demonstrate their knowledge of the mix design procedures by conducting a full mix design for DOT

Table 5 - *Typical Topics Covered for Marshall Method of Mix Design Course*

Topic Title	Day	Approximate Time (Hours)
Responsibilities of the Technician	Monday	½
Aggregates: Physical Properties Sieve Analysis Testing Procedures Specific Gravities Sampling Blending & Batching	Monday	4.5
Binder: PG Grading Physical Properties Sampling Safety	Monday	1
Introduction to Marshall Method of Mix Design	Tuesday	3
SC DOT Specifications	Tuesday	2
Moisture Susceptibility, Rice Gravity, Air Voids, %VMA, %VFA	Wednesday	5
Indirect Tensile Strength Testing and Tensile Strength Ratio Calculations	Thursday	4
Written Exit Exam (Open Book and Notes)	Thursday	4

officials. They must spend three days at the SC DOT's Central Laboratory and conduct the mix design. After this process has been completed they will be certified at Level 2.

Level 2S: Superpave Mix Design Course

The attendees for this course must be certified at Levels 1 and 2. In this course, the concept of SHRP and Superpave mix design are discussed in detail. There is a maximum of ten attendees. Each group (2 or 3 people) conducts a complete mix design. The Superpave mix design procedures developed by SC DOT are covered. This class, like Levels 1 and 2, is very hands-on. The majority of the participants' time is spent in the laboratory. At the end of the week, they must pass a written exam (approximately 3 to 4 hours). The passing grade is 80%.

Level 3: QA/QC Manager Course

This course is designed for people who are in a managerial position. They must be familiar with all aspects of an asphalt operation from selection of the materials, mix design to the troubleshooting of asphalt plants. The course is offered twice per year and a maximum of twelve people are allowed to register. Most of the course is problem solving and troubleshooting. The participants are divided into four groups and they are given several actual cases to solve related to aggregates, asphalt plants, mix design, and statistics. This course does not have any laboratory experiments. Several experienced industry experts are invited to cover topics such as aggregates, drum and batch plants, and specifications. Table 6 shows some typical topics covered in this class. The participants must pass a written exam (3 to 4 hours) to be certified at this level. The passing grade is 85%.

Table 6 - *Typical Topics Covered for QA/QC Manager Course*

Topic Title	Day	Approximate Time (Hours)
Responsibilities of the QA/QC Manager	Monday	½

Aggregates:
    Physical Properties
    Sampling
    Testing Procedures
    0.45 Power Curve
    Blending
    Problem Solving
	Monday	3

Binder & HMA Mixtures:
    PG Grading
    SHRP
    Marshall & Superpave Mix Designs
    Sampling
    Safety
    Polymers
    Problem Solving
	Monday	2.5

Drum Mix Asphalt Plants Batch Plants	Tuesday	5
SC DOT Specifications	Tuesday	2
Data Analysis (Statistics)	Wednesday	7
Written Exit Exam (Open Book and Notes)	Thursday	4

Summary and Conclusions

The SC DOT, in 1993, initiated a program for the certification of technicians involved with asphalt mixtures. There are now five courses (levels). Most of the courses have an enrollment limit of twelve people, therefore, enables the attendees to participate in the discussions. In addition, this allows everyone to conduct a portion of each experiment in the laboratory. Most of the courses include hands-on experiments. All of the courses have written exams. In addition, two of the courses have an oral exam. The technicians that have been certified at the mix design level must also be AMRL certified. The field laboratories must be evaluated and certified by the DOT officials. Most of the courses are team-taught. The classes are reviewed and evaluated at least once per year by the Certification Board. The members of this Board consist of people from industry, DOT officials, and the author. Individuals from industry, consultants, DOT, and academia are responsible for covering a portion of each class. Based on the comments from the participants, many from other states, the courses have been, in general, a welcome addition to the industry.